Developments in Animal and Veterinary Sciences, 17

Safety and quality in food

Developments in Animal and Veterinary Sciences, 17

Safety and quality in food

Proceedings of a DSA symposium 'Wholesome Food for All. Views of the Animal Health Industries' held in Brussels, 29—30 March 1984

Edited by DSA (Bureau Européen d'Information pour le Développement de la Santé Animale), Avenue de l'Yser 19, B-1040 Brussels, Belgium

ELSEVIER

Amsterdam — Oxford — New York — Tokyo   1984

ELSEVIER SCIENCE PUBLISHERS B.V.
Molenwerf 1
P.O. Box 211, 1000 AE Amsterdam, The Netherlands

Distributors for the United States and Canada:

ELSEVIER SCIENCE PUBLISHING COMPANY INC.
52, Vanderbilt Avenue
New York, NY 10017

ISBN 0-444-42409-1 (Vol. 17)
ISBN 0-444-41703-6 (Series)

Printed in The Netherlands

<u>D.S.A. - BUREAU EUROPEEN D'INFORMATION POUR LE DEVELOPPEMENT</u>

<u>DE LA SANTE ANIMALE,</u>

Association Internationale.

D.S.A. is an association of leading manufacturers of animal health products, national and multi-national, and registered under Belgian Law in Brussels. D.S.A. has an established office at 19 avenue de l'Yser, 1040 Brussels.

Its activities are concerned with and confined to all aspects of products to ensure that healthy animals are used in economic agricultural production for human consumption and companion or pet animals, such as dogs and cats, in Europe.

Although acting purely in an advisory and consultative role, D.S.A. functions to educate and inform all sections of the community that modern aids to animal production are safe, wholesome, efficacious and beneficial to the animal and the consumer. To achieve this objective, D.S.A. will carry on active dialogue and publicity with the many interested bodies including general public, regulatory and public health bodies, veterinarians, farmers, food and feed processors, universities, advisory bodies and legislative organisations. All the information conveyed to such people will be based on sound technical and ethical concepts and presented in a comprehensible and relevant manner.

Le D.S.A. est une association de sociétés pharmaceutiques vétérinaires nationales et multinationales. Régi par les lois belges, le D.S.A. a son siège social au 19 avenue de l'Yser à 1040 Bruxelles.

Ses activités sont principalement et exclusivement axées sur tous les aspects des produits pour la santé animale afin d'assurer l'utilisation d'animaux sains en production animale destinée à la consommation humaine ainsi que la santé des animaux de compagnie (chiens, chats) en Europe.

Bien qu'il n'agisse qu'à titre consultatif, le D.S.A. a pour objectif d'éduquer et d'informer tous les secteurs de la communauté sur la sécurité et l'efficacité des techniques modernes de production animale et sur les bénéfices qu'entraîne leur utilisation. Dans ce but, le D.S.A. maintiendra un dialogue actif avec les diverses parties intéressées, et notamment le grand public, les autorités législatives et exécutives responsables de l'enregistrement de ces produits et de la santé publique, les vétérinaires, producteurs agricoles, fabricants d'aliments pour la consommation humaine et fabricants d'aliments pour animaux, universités, et autres organes consultatifs. Toute information transmise à ces personnes ou organes sera fondée sur des données techniques et concepts éthiques sains et présentée de manière compréhensible et appropriée.

Die D.S.A. ist eine Vereinigung von nationalen und multinationalen Herstellern von Produkten für die Erhaltung der Tiergesundheit. Die D.S.A. ist in Belgien registriert und hat ihr Büro in der 19 avenue de l'Yser, 1040 Brüssel.

Die D.S.A. beschäftigt sich mit allen Gesichtspunkten von Produkten, die die Gesundheit von Tieren zur Nahrungsmittelproduktion sowie von Luxustieren in Europa gewährleisten.

Obwohl die D.S.A. eine beratende Funktion hat, sieht sie vor allem ihre Aufgabe darin, alle Bereiche der Europäischen Gemeinschaft über die Sicherheit und Wirksamkeit moderner Tiererzeugungstechniken und über die Vorteile ihrer Anwendung zu informieren. Deswegen unterhält die D.S.A. einen aktiven Dialog mit allen betroffenen Parteien, und zwar der Öffentlichkeit, den Behörden, die für die Zulassung von Produkten und die öffentliche Gesundheit verantwortlich sind, den Tierärzten und Landwirten, der Futter- und Nahrungsmittelindustrie, den Universitäten und anderen beratenden Organisationen. Alle Informationen an diese Gesprächspartner basieren auf eindeutigen wissenschaftlich erarbeiteten Ergebnissen und den ethischen Grundvorstellungen unserer Gesellschaft. Die D.S.A. bemüht sich, all ihre Mitteilungen in einer veständlichen, dem jeweiligen Partner angepassten form zu verfassen.

CONTENTS

Dr. Jean-Claude Bouffault
Roussel-Uclaf, France,
President of D.S.A. - Bureau Européen d'Information pour le Développement
de la Santé Animale, association internationale, Brussels.

FOREWORD

Ladies and Gentlemen,

It is an honour to welcome you to the first symposium organised by
D.S.A. on food of animal origin, which is both a daily necessity and one of
the pleasures of life, and its relationships with modern technology,
industrial livestock farming, its benefits, constraints and the obstacles it
has to overcome.

A permanent ordeal, kept ongoing by the media, is made against the
whole of mordern animal production techniques : for our British friends, the
criticism is oriented towards environment, conditions of housing, living and
slaughter of the production animals; our Latin friends question the taste of
animal production and require in particular perfect safety for the consumers
of these products. A last argument is now opposed to us : the excessively
high quantity of food products available.

It is necessary to look back into the subconscious of human soul to
understand these attacks. Yesterday, collective faith and custom helped man
to accept his destiny. Hunger and poverty focused on immediate problems which
hid existential problems. Today, material progress and science have broken
those protective shells. Today's world is worried and nervous. Hence a
return to summary instincts and the deification of one entity : nature, a
rigid dogma. Polluters are accused. Natural purity is protected. The
irrational triumphs with a lack of lucidity and subjectivity.

Myths are errors of human brain which are erased by knowledge, but
this is a very slow process : four centuries to accept the fact "microbe" ...
The trial made by our civilisation to science and technology seems to come out
of a deep trouble. Irrational beliefs continue to spread although conditions
of life bring a sanitary state and a life-span unknown till now. Our era is
characterized by an acute sense of the critical. Everything is questioned and
phantasies have changed, just like psychologists have replaced confessors
after the collapse of faith ... but not for the same fees !

In his book "Le Refus du Réel" (the rejection of reality) (Ed. Robert
Laffont), Prof. Tubiana says that man expresses his personality by what he
eats and even more by what he refuses to eat. It is thus not surprising that
food problems should be one of the first choice fields of our modern myths.
He goes on by saying that from a medical point of view, the fear of chicken
with hormones and antibiotics in animal feedingstuffs makes no sense.

The use of these antibiotics in animal health is criticized, but no effort is made to avoid excesses in human medicine which is characterized by a ten times higher consumption. The attention is drawn to minor problems, the presence of colouring agents in yoghurts and sweets hide major problems such as excess of fats and sugars which cause stoutness, tobacco ...

Etymologically, artificial means man-made. Nothing is more natural than the molds on a badly preserved meat from a sick animal and nothing is more carcinogenic than a mycotoxin. The safety of food supplied by so-called industrial animal farming is explained by the absolute need to maintain health conditions on farms with high animal concentrations. The rigorous control of products is a further insurance.

There is a world of difference between the weekly chicken hoped for by the Prime Minister of François Ier (French King, XVth-XVIth century) and today's daily fast-food. We are far from the enormous meals alternated with periods of hunger of mankind's early days. Cereal-based meals at regular times have followed upon these. Since the last world war, our daily food is based on more animal products. We have long forgotten hunger and caprices of the weather. Chicken has become a daily food, veal and eggs remain at a reasonable price as a result of so-called industrial production.

This is an important sociological point : this industry enables the small farm owners, Europe's gardeners, to survive with the aid of their governments and maintain the environment. Europe has a beautiful garden which it must maintain. It is a luxury, but the quality of our life is the price.

It is essential to pass this message. Nothing is more complex and more indispensable than information at all stages of life and in its broader sense :
- at the level of the cell and its specific receptors,
- at the level of the organism, in which many millions of messages are passed simultaneously, in particular by hormones.

Man is immersed in a sea of messages and sensations which determine his personality and condition his actions, says Prof. Tubiana. But man takes note of the type of information he wants to hear and ignores the rest. Now a dead or unobedient cell will be the starting point of a cancer, just as pre-conceived ideas will block information and cause disorders which will be harmful for the essential harmony of society.

A complex reactivity, the need for redundancy, even with language that has replaced the primitive signals exchanged among animals, condition this receptivity.

Finally, dialogue should be preferred to one-way information (the article in the newspaper, the speech on the radio or television information).

Hence the need for a forum such as this one, in which you are participating.

Dr. Jean-Claude BOUFFAULT

President of D.S.A.

SESSION I: ANIMAL PRODUCTION IN WESTERN EUROPE

CONTRIBUTORS

Mr. Yves Domzalski, B.E.U.C., Brussels.

Dr. Graham Meadows,
Agricultural Advisor to the President of the European Commission.

Prof. Giovanni Ballarini,
Director of the Clinica Medica Veterinaria of the University of Parma (Italy),
Member of the Scientific Committee for Animal Nutrition of the European
Communities.

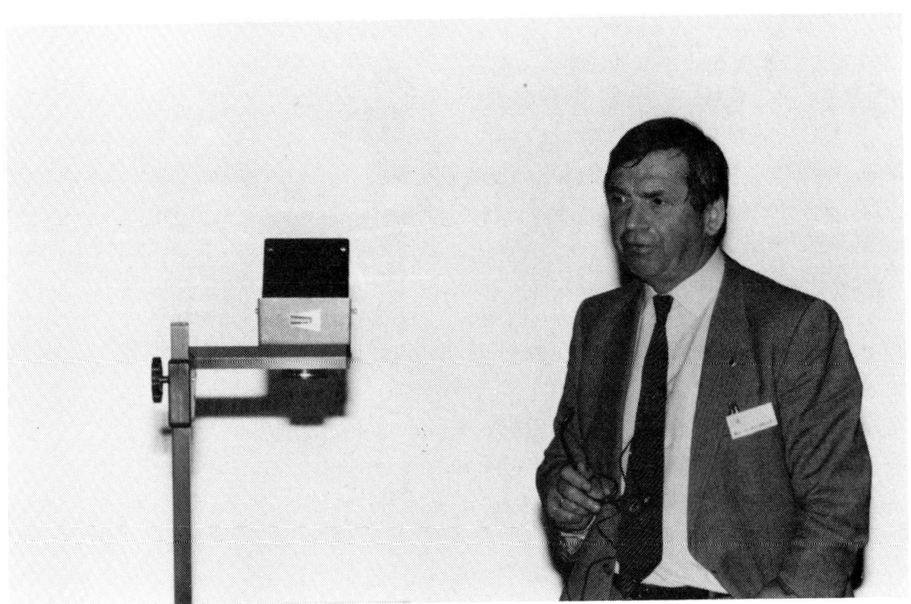

Prof. Marcel Vanbelle,
Faculté des Sciences Agronomiques, Laboratoire de Biochimie de la Nutrition,
U.C.L.-Louvain-La-Neuve (Belgium),
Member of the Scientific Committee for Animal Nutrition of the European
Communities.

Dr. Robert Marsboom,
Janssen Pharmaceutica, 2340 Beerse, Belgium.

<u>FOR A TRUE COMMON ALIMENTARY POLICY</u>

Yves DOMZALSKI,

B.E.U.C., Brussels

The Bureau Européen des Unions de Consommateurs (BEUC) which has its statutory office in Brussels represents at the level of the Community's Institutions the interests of fifteen member consumers' associations of countries of the European Communities.

The BEUC and its members associations consider that, when national or community policies and decisions involve various interests, these must reflect the whole of these interests : producers, of course, but also transformation industry, distributors, consumers, etc.

This requirement corresponds to one of the targets of the consumer associations : to achieve recognition of their right to be considered as true partners as well as all the other economical sectors.

Progress has certainly been made in this respect over the last fifteen years in some Member States of the E.C. It is also true at Community level since a number of years. However, it must be recognized that consumers continue to be considered at best as a sort of "necessary disease" at the end of the economic chain, mostly as troublemakers whom one could do without, particularly in today's times of economical crisis.

If one tries nonetheless to overcome this concept, it will be appreciated that the consumers' demands, rather than being systematic shackles for the producers, may also present means of improving their production and consolidating the market. It will also be noticed that these demands can contribute to the resorption of some economical imbalance or disorders, for example regarding protectionism or inflation.

One area where consumers are considered as troublemakers, whereas some of their demands would precisely help avoid economic nonsense and improve product quality, is that of the Common Agricultural Policy and production. For several years, consumers' associations have been asking that the "Common Agricultural Policy" be transformed into a "Food Policy" which, at present, is lacking in the E.C., although the Commission has taken the idea on its own account last year.

The Common Agricultural Policy until now is limited to salary negotiations with producers and to the granting of guaranteed prices which defy any economic balance. It is urgent to initiate a common policy of food which, on one hand, would reckon not only with the economic interests of producers, but also with those of all concerned parties - amongst whom the consumers - and, on the other hand, would permit a coherent and effective improvement of health and quality standards of products.

1.   The C.A.P. and the economic interests of consumers

It is usually said that the C.A.P. is victim of its success. It would be more appropriate to say that it is victim of its non-respect of the targets fixed by the Treaty of Rome. Indeed, Article 39 of the E.C. Treaty mentions that "the C.A.P. has for objective to increase agricultural productivity ... by ensuring the rational development of production (...) to stabilize the markets (...) to ensure reasonable price of supplies to the consumers".

At the beginning of this year there was a stock of over one million tonnes of skimmed milk, 900,000 tonnes of butter, 420,000 tonnes of beef meat, plus stocks of sugar, wine, cereals, fruit and vegetables; 70 % of Greek apricot harvest have been destroyed in 1983. The recent agreement of the 12th March 1984 on the reduction of milk production is presented as a success : in fact, production which achieved 103 million tonnes in 1983 should be limited to 98.8 million tonnes in 1984-85.

It is unacceptable that, in spite of these surpluses, the consumers continue to pay increasingly artificial prices for basic products such as milk, butter, sugar or beef meat and the multitude of food products that are derived from these.

For your information, Europeans spend on their food - beverages excluded - over 20 % of the total family budget, with wide variances amongst Member States (from 20 % in Germany to 42 % in Eire).

The consumer is doubly penalized, since as a tax-payer he must also sponsor the stocking and export of surpluses. The cost of the CAP is about 70 % of the Community's budget, which means an annual expense of about BEF 9,400 for a four persons' family.

Pierre Lelong, President of the E.C. Finance Department said : "The agricultural markets are organised in a fully incoherent manner. Massive savings could be made in the Community's budget, but they would be unpleasant for everybody."

Experience indeed shows that surplus production at guaranteed prices not only brings profit to the large agricultural producers, but also to various sectors involved in the agro-food chain. The aberrant example of the feeding of calves in batteries in well known. If they were fed directly with milk, the calves produced in the E.C. would absorb 10 % of the Community's milk production, which would save 8.3 billion FF of credits to exportation.

The dairy and feedingstuffs industries have broken this short circuit to sell their replacement feeds, apparently competitive with milk as a result of the Community's credits and premiums.

Another significant example is that of beef meat. Under the heading "Beef" of Doc. 83/500 of July 29, 1983, the E.C. Commission "is worried by the risk of future imbalance which could characterize this sector of the market". Indeed consumption remains stable whereas production increases regularly.

As a consequence, the public stock at the beginning of January 1984 was in the order of 420 to 430 thousand tonnes. It has cost the E.C. 460 million ECU in 1983 (i.e. 3.17 billion FF, 21.6 billion BEF or 1.03 billion DM). These figures clearly indicate that it would be nonsense to try to increase production further. Now, the E.C. Commission, under the pressure of various private interests, wishes to legalise a number of growth hormones !

These economic aberrations are possible because the legitimate interests of consumers have remained neglected until now. To reverse this trend, the BEUC demands in particular to promote a reform of the CAP which is based namely on three principles formulated by the Commission itself in 1981 :
- a price policy based on a shrinking of the difference between the Community prices and the prices of its competitors;
- a reduction of the unlimited price guarantee and the automatic purchase of surpluses at the expense of the long term objectives of the consumption and the production;
- that the price policy should no longer be the only way to respond to the demands of farmers concerning their revenue and that it should be possible to set up a system of direct aids for some farmers based on specific conditions.

2.   C.A.P. and the quality of food

The Treaty of Rome has not foreseen to assign to the C.A.P. the function of improving health and quality standards of food products. As consumers' associations have been demanding for quite some time, the Commission has recognized less than a year ago that "in the present economic conditions, a common agricultural policy is not conceivable outside the larger concept of food policy" (in "Nouvelles orientations pour le Développement de la PAC, Juin 1983, COM/83/320). But is it clear for the Commission that this concept of food policy incorporates the issues on health and quality ?

Of course the Community issues laws in the area of food, but its action mainly consists of harmonising national standards - most often based on the smallest common denominator - and not of promoting significantly the quality of food. In view namely of the evolution of production techniques and nutritional habits, the role of food vis à vis the health of the population and the increasing importance of the agro-industrial sector in our economy it is however important to study and carry out some research in all the areas related to food and to draw conclusions of it from the legislative point of view. In short a policy which would improve the health standards and the quality of food products in the whole of the EC countries is still to be created.

The basic requirement for such a policy to function effectively is to ensure a real transparency of information and scientific research and to take into account all interests involved. In this respect, two important dossiers which are currently reviewed by the Community authorities represent examples that should not be followed : growth hormones for livestock production and American wines. It is unacceptable that elementary rules of concertation should not be respected and that access to information should be restricted in order to protect industrial or non-European countries' interests.

The first task of a food policy must be to permit the development of research, particularly in the field of nutrition, epidemiology and toxicology (basic research on analytical methods for toxicity evaluation, study of metabolism and toxicity mechanisms of some substances, mutagenesis, carcinogenesis, long term toxicology at low doses and associations of toxic compounds).

It must also favour reflexion and research on the role and the influence of technological innovations, nutritional behaviour and consumption models. How can for example innovation better be adapted to needs ? How to measure the need for the general use of some substances ? How to define concretely the quality, beyond theoretical concepts ?

To initiate a food policy also means that information and education are substantially developed : these are two decisive elements of individual and collective choice which permit the improvement of consumers' participation in the choice of a food policy.

Once the elements permitting the appreciation of the problems with a maximum of objectivity and transparency are gathered, a coherent regulation must be prepared. We will only emphasize two points here : firstly the need for the legislative authority to define the notion of risk or safety of the substances and processes, and secondly the need to foresee a permanent adaptation of legislation with the improvement of scientific knowledge.

Conclusion

It is necessary to permit a reflexion and a discussion between the agro-food industry, researchers, farmers and consumers in order to link better production and technological innovation to the actual needs of the population.

The problem is that of the mastery of the evolution of our nutrition. Will it be increasingly imposed by a number of private interests, mainly based on criteria of economical profitability, or will it result from consultation where the free choice of consumers will also be taken into account ?

The promotion of a common food policy is the logical extension of the C.A.P. It is the task of the E.C. to coordinate public and private research, to set up adequate consultation and decision structures and progressively to prepare a common legislation which goes beyond the minimum harmonisation so that Europeans become the masters of their nutrition.

RESUME

POUR UNE VERITABLE POLITIQUE COMMUNAUTAIRE DE L'ALIMENTATION

Yves Domzalski
B.E.U.C., Bruxelles

Le Bureau Européen des Unions de Consommateurs (BEUC) représente les intérêts de quinze associations membres des pays de la Communauté Européenne auprès des Institutions Communautaires. Il considère que, lorsque des politiques et des décisions nationales et communautaires mettent en jeu des intérêts divers, celles-ci doivent refléter l'ensemble des intérêts concernés: producteurs, transformateurs, distributeurs, consommateurs... Cette exigence correspond à l'un des buts poursuivis par les associations de consommateurs : faire reconnaître leur droit à être considérés comme de véritables partenaires, au même titre que l'ensemble des autres secteurs du jeu économique.

Certes, des progrès ont été réalisés à cet égard depuis une quinzaine d'années, tant au niveau de certains Etats Membres que de la Communauté. Néanmoins, force est de reconnaître que les consommateurs continuent à être considérés comme une sorte de "mal nécessaire", surtout en cette période de crise économique. Alors que certaines de leurs demandes permettraient d'éviter des absurdités économiques et d'améliorer la qualité des produits, un domaine où les consommateurs passent pour des gêneurs est celui de la politique et de la production agricoles communautaires. Depuis plusieurs années les associations de consommateurs réclament que la "Politique agricole commune" (PAC) se transforme en une "politique de l'alimentation".

Il est nécessaire de permettre une réflexion et une discussion entre industriels de l'agro-alimentaire, chercheurs, agriculteurs et consommateurs afin de mieux subordonner la production et l'innovation technologique aux besoins réels de la population. Le problème posé est celui de la maîtrise de l'évolution de notre alimentation. Celle-ci sera-t-elle de plus en plus imposée par un ensemble d'intérêts privés, principalement en fonction de critères de rentabilité économique, ou résultera-t-elle d'une concertation où le libre choix des consommateurs entrera aussi en ligne de compte ?

La promotion d'une politique alimentaire communautaire constitue le prolongement logique de la PAC. Il appartient à la CE de coordonner la recherche publique et privée, de mettre en place les structures de concertation et de décision adéquates et de mettre progressivement en oeuvre une législation communautaire qui dépasse le cadre de l'harmonisation minimale afin que les européens deviennent maîtres de leur alimentation.

ZUSAMMENFASSUNG

FÜR EINE ECHTE GEMEINSCHAFTLICHE ERNÄHRUNGSPOLITIK
Yves Domzalski
B.E.U.C., Brüssel

Das Europäische Büro der Verbraucherverbände (BEUC) nimmt die Interesse von fünfzehn Mitgliedsverbänden aus Ländern der Europäischen Gemeinschaft wahr. Es ist der Meinung, daß dann wenn unterschiedliche Interesse
durch gemeinschaftliche und einzelstaatliche Entscheidungen und Politiken
berührt werden, diese die Gesamtheit der betreffenden Interessen zu berücksichtigen haben : Erzeuger, Weiterverarbeitung, Vertrieb, Verbraucher ...
Diese Forderung entspricht einer der Zielsetzungen der Verbraucherverbände :
Durchsetzen der Anerkennung ihres Rechts, als echte Partner wie alle übrigen
Sektoren des wirtschaftlichen Zusammenspiels betrachtet zu werden.

Sowohl in bestimmten Mitgliedstaaten als auch auf Gemeinschaftsebene
sind zwar seit etwa fünfzehn Jahren in dieser Beziehung Fortschritte erzielt
worden, aber wir müssen gestehen, daß Verbraucher weiterhin als eine Art
"notwendiges Übel" gelten, besonders in der heutigen Wirtschaftskrise. Obwohl
die Erfüllung bestimmter Forderungen der Verbraucher wirtschaftliche Absurdi-
täten verhindern und die Qualität der Erzeugnisse verbessern könnte, sind
gerade die gemeinschaftliche Agrarpolitik und -produktion ein Bereich, in dem
die Verbraucher als Störenfriede gelten. Seit mehreren Jahren fordern die
Verbraucherverbände, die "Gemeinschaftliche Agrarpolitik" (GAP) müsse in eine
"Ernährungspolitik" umgewandelt werden.

Notwendig ist das Ermöglichen von Überlegungen und Erörterungen
zwischen Verantwortlichen der Landwirtschaft und Lebensmittelindustrie, Forschern, Landwirten und Verbrauchern im Hinblick auf eine bessere Unterordnung
der Produktion und technologischen Innovation gegenüber den echten Bedürfnissen
der Bevölkerung. Es geht um das Problem des Beherrschens der weiteren Entwicklung unserer Ernährung. Wird sie uns in zunehmendem Umfang von einer Gesamtheit von Privatinteressen vorgeschrieben, und zwar hauptsächlich nach wirtschaftlichen Rentabilitätskriterien, oder soll sie sich aus einer Konzertation
ergeben, bei der auch die freie Wahl der Verbraucher mit in Betracht gezogen
wird ?

Die Förderung einer gemeinschaftlichen Ernährungspolitik ist der
logische Fortsatz der GAP. Der EG obliegt es, die öffentliche und die private
Forschung zu koordinieren, angemessene Konzertations- und Entscheidungsstruk-
turen einzuführen und stufenweise gemeinschaftliche Rechtsvorschriften zu ver-
wirklichen, die den Rahmen der minimalen Angleichung überschreiten, damit die
Europäer ihre Ernährung in den Griff bekommen.

THE PHILOSOPHY OF THE AGRICULTURAL POLICY

Graham MEADOWS,

Agricultural Advisor to the President of the European Commission

The Common Agricultural Policy is at a turning point. Created 25 years ago partly as a means of builduing up the European Community's potential for food production and thus of assuring food supplies, the policy is on the point of limiting the production of a major farm product, milk, through quotas. The decision, so far only nine-tenth taken, will mark the most important single change since the policy's inception in the late 1950's : acceptance that, for agricultural output, you can have too much of a good thing; that production beyond a particular threshold can only be permitted if the farmers themselves bear the disposal cost of excess products.

In these remarks I would like, because of the importance of this change and of current negotiations, to dwell on the outlook for the Common Agricultural Policy, something which is of vital concern.
- It is the Common Agricultural Policy which assures food supplies for the Community's 270 million inhabitants (soon to be 320 million when Spain and Portugal join the Community).
- It is the Common Agricultural Policy which assures the economic framework in which the Community's 8 million farmers operate (11.5 million after enlargement).
- It is the Common Agricultural Policy, and in particular its future shape, which lies at the heart of the present negotiation for the relaunch of the European Community.

First, I would like to refresh your memories concerning the extent of the present negotiations. Second, we can look in some detail at the problems inside the agricultural sector that justify the adaptation of the Common Agricultural Policy. Third, we can draw conclusions regarding the future policy framework for livestock markets.

1.   The extent of the present negotiations

The present negotiations turn around a number of elements, which together form an interconnected package.

One concerns the Community budget. Community spending is financed from customs duties and agricultural levies collected on Community imports and, for the moment, the first 1 % of national VAT receipts. The Community is now at the limit of these resources and the Common Market Commission has proposed that the present ceiling on the VAT payments should be raised.

The Community's Heads of State and Government reached a conditional consensus to lift the ceiling on present income from VAT to 1.4 %. The Commission says this is not enough - especially in the light of the Community's enlargement to include Spain and Portugal - and that such an agreement would condemn the Community to a continuing preoccupation with its finances instead of allowing it to devote its main energies to the more important questions of stemming industrial decline, reducing unemployment and promoting development.

Another factor within the budgetary field is the lack of balance between some Member States' contributions, their relative wealth and their ability to benefit from existing Community policies. This is characterised as the British budget problem and the search here is for a corrective mechanism rather than an annual wrangle about the size of ad-hoc payments.

Heads of State and Government and their Foreign Ministers got tantalisingly close to settling this question. Finance and Foreign Ministers were going to look at it again. There remains, however, a danger that the United Kingdom and the other Member States will drift further apart with the result that some time will elapse before conditions are again opportune for agreement.

One other budgetary question concerns the acceptance of "greater discipline" - and here we begin to approach the problem of the Common Agricultural Policy. The call for a Community budget discipline originates in a desire to subject Community spending to the same constraints as those felt by Member States nationally. Heads of State and Government were close to agreeing that the rate of growth of agricultural expenditure, calculated in a particular way, should be less than the rate of growth of potential own resources.

This in turn implies considerable adaptations to the Common Agricultural Policy. Agriculture Ministers of the Ten have almost agreed a series of adaptations for different market support systems and creating a system for the dismantling of frontier payments called Monetary Compensatory Amounts. Some difficulties remain, most notably Irish reluctance to accept a quota system for milk production, and Agriculture Ministers met again to try to find a complete settlement. Obviously, policy adaptations are vitally necessary if the engagement to limit the rate of growth of agricultural expenditure is to have a real meaning.

Even this list does not cover the full scope of the negotiations. There is a consensus on a many-sided programme for new initiatives in industrial policy, and for better coordination of structural funds, including the adoption of special economic packages for Mediterranean regions.

Let me conclude this partial survey of what is now being negotiated by emphasising those parts that are more directly tied to the question of the future of the Common Agricultural Policy. Principal element is the need to increase the financial resources of the Community. Some Member States make their agreement to new financial resources conditional on the control of agricultural expenditure. This linkage shows up in the consensus concerning the need for greater budget discipline and adaptations to the Common Agricultural Policy.

2.   Problems inside the agricultural sector

In any discussion about the Common Agricultural Policy and its problems it is necessary to bear in mind the political and economic background. We have just looked at the political background. Now let us consider briefly some economic factors.

The Community is still gripped by an economic recession that began to make itself felt ten years ago. Its depth can be gauged by the fact that about 12 million people are without work and about a third of those without jobs are under the age of 25.

Agriculture has largely escaped the worst effects of this recession, partly by its own efforts and partly due to the Common Agricultural Policy. It slimmed its workforce considerably in the 1960's, it has continued its never-ending effort of modernisation, it has benefitted from the rapid growth of the food processing sector.

Sustained and sustainable recovery from this recession demands that the Community economy be restructured, the size of some traditionally important sectors being reduced. The Community steel sector, for example, has had to reduce employment by a third (300,000 people) and must go further. This naturally affects the climate of opinion when one comes to consider the future shape of the Common Agricultural Policy. Agriculture cannot stand aside from this economic restructuring.

We can summarize a number of other factors.
-       The Community's 8 million farmers are distributed unevenly between Member States with marked variations in their productivity. In the United Kingdom, farmers and farmworkers together account for about 3 % of the workforce and produce 2 % of national income. In Greece, 30 % of the workforce, 17 % of national income.
-       Farmers in different Member States face considerably different inflation rates; the present range, for instance, is between 2.3 % in Germany and

about 17 % in Greece. In part these differences are compensated within the Common Agricultural Policy by the evolution of currency exchange rates. Even so, considerable problems remain for a common pricing policy.

- Farm income vary widely between region and type of enterprise. Generally speaking, the bigger the farm, the bigger the income (measured in net value added); and, remember, farm size varies considerably between Member States; cereal produce bigger incomes than milk, milk than meat. But I underline the qualification "generally".

So much for the economic and political background, what are the problems ?

- First there are market disequilibria. The normal indicator cited to demonstrate the depth of these market difficulties is the stock level of key products. Butter and skimmed milk powder stocks are not far short of one million tonnes, beef stocks are about 400,000 tonnes, wheat stocks are 6 million tonnes and are falling slightly, olive oil stocks are increasing and are expected to reach 200,000 tonnes.

- Second, farm incomes pose problems. The gap between farm sector income and that of the general economy has widened since the mid-1970's. Despite the increased production that has given rise to present high stocks, real farm income is now lower than it was in 1975 while that in the rest of the economy is higher.

- Third, there is the development of agricultural expenditure. After a period at the beginning of the 1980's when agricultural support expenditure decreased not only as a proportion of the Community budget, but also in absolute terms, it is again growing rapidly. This growth in agricultural support expenditure is all the more worrying since it is taking place at a time when commodity stocks are increasing. It is normally - because of Community accounting procedures - the subsidised disposal of stocks which leads to high budget expenditure.

- Fourth, there is the Community budget. I have already said that the Community is at the limit of its financial resources. Indeed, in 1984, it is already clear that the Community will be unable to finance the agricultural policy out of its normal budgetary resources. Some other means will have to be found and the need for temporary, extra financing can be expected to continue until Member States agree to furnish the Community with more budget resources, at the earliest the beginning of 1986.

These factors have together convinced all sections of opinion that the Common Agricultural Policy needs to be remodelled. Different people, however, draw different policy conclusions. Some can be more quickly rejected than others. For example, it is often suggested that it is a scandal that the Community should be talking of reducing food output in a world in which many are starving. Policy conclusion : increase Community food aid and give away surplus production to countries which need food.

This conclusion overlooks a number of other factors. Countries which are short of food would often rather produce it themselves; they are short of specific commodities which are often not the product that the Community has in surplus; one cannot build a food aid policy on excess in domestic agricultural policy; the taxpayer cannot be expected to increase his financing of the agricultural sector in order that food can be given away, no matter how deserving the recipient.

Another suggestion is that overproduction is due to the use of 20th century technology. For example, I have heard it said, though not at first hand, that present beef surpluses are due to the use of growth stimulating products. Policy conclusion : slow down the uptake of technological development and thus reduce production.

The conclusion again overlooks a number of factors. The level of agricultural incomes in relation to the rest of the economy is a function of the lower rate of productivity in agriculture compared to other sectors. Some sort of control on technological uptake - whether it be in the form of preventing the use of a new machine, or of a new variety of seed, or of a growth stimulant which had fulfilled necessary criteria in terms of product safety - is merely a further brake on the development of farm income. The Community draws a clear distinction between the need to control the growth of production and the need to maximise productivity : in the milk sector, for example, as soon as the quota scheme is in place we shall make Community subsidies available to milk producers so that they can modernise their farms.

Then one comes to conclusions which are less easy to reject. For example it is often said that overproduction and high expenditure are due to prices that are too high. Policy conclusion : reduce prices and so increase consumption, and put a brake on production.

Another example in this category, again it is said that the Community lacks aggression in selling to world markets and demonstrates an inexplicable desire to import products the Community does not need. Policy conclusion : export more, import less.

In both of these conclusions, there is more than a grain of truth. In seeking to adapt its market support systems to changed conditions, the Community has taken this sort of view into account : balancing its market objectives against the needs of farm incomes and the availability of budget finance.

3.   <u>Future policy for livestock products</u>

Let us look at three livestock sectors, milk, beef, and pig and poultry meat. In all three cases, the considerations discussed above lead to a different policy mix.

In the milk sector, policy for the next five years will be built on a quota system. Farmers will be able to produce up to a given quantity, more or less equal to their production in 1981; production beyond that level will be charged a production levy equal to either 100 % or 75 % of the milk price depending on the way the quota is applied. Already, nine Member States have accepted this scheme - which would come into force on April 1, next Sunday, only Ireland is holding up final agreement.

The fact that the full support price is paid for production within the quota will ensure the income needs of producers. Production levels will be controlled by the quota system so that - although producers will still be able to sell to the intervention authorities at the official support price - the level of product stocks will diminish and budget expenditure will be reduced. When the system is working effectively, the support price for milk will be free to increase.

In the beef market the more traditional market support system will continue to operate. Here, however, it will be refined so that support purchasing will be limited to different parts of the carcases at different times of the year. In other words, we shall move even further away from the system we applied until a few years ago - that producers could sell whole carcases to the intervention authority at no matter what time regardless of the prevailing market price. In addition we shall remove a number of different production subsidies.

For poultry and pig meat we shall contiue as at present - these two industries will continue to operate in almost free market conditions with minimal price support. In fact, one sometimes has the impression that the major effect of agricultural policy on these sectors is in the form of fall-out from other areas. For example, we limit the import of cereal substitutes as part of our cereals sector policy, pig and poultry input costs rise; currencies are realigned with corresponding increases in Monetary Compensatory Amounts, pig producers find themselves competing against what are to all intents and purposes subsidised imports; the Community introduces an unlimited deficiency payment for lamb, then pig and poultry producers in the same Member States find demand for their product shrinking.

4. <u>Conclusion : the turning point in the Common Agricultural Policy</u>

How does all this add up to a turning point in the Common Agricultural Policy ?

The first point to make in conclusion is that the Common Agricultural Policy has been and is successful. It has allowed the Community agriculture sector to exploit its potential, it has provided the economic climate that has allowed the adoption of new production techniques, it has helped to drastically change the structure of the industry. In these ways it has helped to assure food supplies at reasonable prices for Community consumers.

But quite clearly any policy will periodically require adjustment. In recent years it has become increasingly clear that farmers were continuing to increase output, not because the food was required with the Community, but because market signals had somehow become blocked. The system of guaranteed prices for unlimited quantities told farmers : "Produce, more output is required". It was the taxpayer who received the correct market signal : "Reduce output, normal markets do not exist for present production levels. Output can only be disposed of with ever-increasing subsidies".

Over the two years, policy adaptations have aimed to get the right market signal through to producers. The Community is now on the brink of bringing a much greater awareness of market reality into the milk sector. If his production goes beyond his quota, it will be the producer, not the taxpayer, who will pay the disposal cost of production beyond the quota. The same awareness will grow for beef producers bringing them more closely into line with the pig and poultry sectors where - because of the intensivity of their production - farmers have always had to respond to market reality. The same will also apply to cereal growers and other producers of arable crops.

This does not mean that producers are about to be dumped on the market and left to their own devices. Market prices will still be supported, the tax payer will still have a burden to bear, but support will not be available for an unlimited quantity of product. In this way the Common Agricultural Policy will be adapted to present economic conditions and the agricultural sector will make its contribution to the restructuring of the Community economy.

RESUME

LA PHILOSOPHIE DE LA POLITIQUE AGRICOLE

Graham MEADOWS
Conseiller du Président de la Commission Européenne

La politique agricole commune est à un tournant. Créée, il y a 25 ans, partiellement pour établir le potentiel de production alimentaire de la Communauté Européenne et donc pour assurer la production alimentaire, la politique est arrivée au point de limiter l'un des produits fermiers principaux, le lait, par des quotas de production. La décision, acceptée par neuf des dix membres, représente le changement le plus important depuis la conception de cette politique vers la fin des années 50 : il est accepté qu'en matière agricole, il soit possible d'avoir pléthore d'un bon produit et que la production au-delà d'un seuil déterminé n'est permise que si les fermiers supportent eux-mêmes le coût de l'évacuation des excédents.

Il est certain que la politique agricole commune a été un succès et le reste. Elle a permis au secteur agricole de la communauté d'exploiter son potentiel, a favorisé le développement d'un climat économique permettant l'adoption de nouvelles techniques de production et a aidé à changer de façon impressionnante la structure de l'industrie. Ainsi elle a assuré aux consommateurs de la Communauté la production d'aliments à des prix raisonnables.

Mais il est évident que toute politique doit être ajustée périodiquement. Dans les dernières années, il est apparu que les fermiers continuaient à augmenter leurs productions, non pas parce que la demande s'en faisait sentir, mais parce que les signaux d'alarme du marché ne fonctionnaient plus. Le système de prix garantis pour des quantités illimitées les a incités à accroître leurs productions aux dépens du contribuable qui paie pour l'évacuation des excédents. Pendant les deux dernières années, les adaptations de la politique ont visé à rétablir le signal pour les producteurs. La Communauté tente de faire prendre à nouveau conscience de la réalité du marché laitier. La même prise de conscience sera imposée aux producteurs de viande bovine pour les rendre aussi sensibles aux conditions du marché que les producteurs de volaille et de porc, et aussi aux cultivateurs.

Ceci ne signifie pas que les producteurs vont être lâchés sur le marché et livrés à eux-mêmes. Les prix continueront à être soutenus et le contribuable devra encore en supporter la charge, mais plus pour une quantité illimitée de produits. Ainsi, la politique agricole commune sera adaptée aux conditions économiques actuelles et le secteur agricole pourra contribuer à la restructuration de l'économie communautaire.

ZUSAMMENFASSUNG

DIE PHILOSOPHIE DER AGRARPOLITIK

Graham Meadows
Berater des Präsidenten der Europäischen Kommission.

Die gemeinschaftliche Agrarpolitik steht vor einem Wendepunkt. Diese Politik wurde vor 25 Jahren zum Teil dafür geschaffen, das Potential der Nahrungsmittelproduktion in der Europäischen Gemeinschaft zu nutzen und infolgedessen diese Nahrungsmittelproduktion zu sichern. Jetzt hat sie einen Punkt erreicht, an dem eines der wichtigsten landwirtschaftlichen Erzeugnisse, nämlich die Milch, mit Hilfe von Produktionsquoten eingeschränkt wird. Diese von neun der zehn Mitglieder angenommene Entscheidung ist die bedeutendste Änderung der betreffenden Politik seit ihrer Gestaltung gegen Ende der 50er Jahre : man nimmt hin, daß es im Agrarbereich zu einem Überfluß bei einem guten Erzeugnis kommen kann und die Produktion oberhalb einer bestimmten Schwelle nur gestattet ist, wenn die Landwirte selbst die Kosten einer Beseitigung der Überschüsse tragen.

Sicher ist, daß die gemeinschaftliche Agrarpolitik ein Erfolg war und bleibt. Sie hat dem landwirtschaftlichen Sektor die Möglichkeit gegeben, sein Potential zu nutzen, sie hat die Entwicklung eines wirtschaftlichen Klimas begünstigt, das den Einsatz neuer produktionstechnischer Verfahren ermöglicht und sie hat dazu beigetragen, die Struktur der Industrie auf eindrucksvolle Art und Weise zu verändern. So hat sie für die Verbraucher der Gemeinschaft die Erzeugung von Nahrungsmitteln zu angemessenen Preisen sichergestellt.

Offensichtlich muß jedoch jede Politik in bestimmten Zeitabständen angepasst werden. In den letzten Jahren hat es sich ergeben, daß die Landwirte ihre Produktionen weiter steigerten, und zwar nicht weil eine Nachfrage spürbar wurde, sondern weil die Alarmsignale des Marktes nicht mehr funktionierten. Durch das System der Garantiepreise für unbegrenzte Mengen wurden sie veranlasst, ihre Produktionen auf Kosten des Steuerzahlers zu erhöhen, der für die Beseitigung der Überschüsse bezahlen muß. In den letzten zwei Jahren galten die Anpassungen der Politik der Wiederherstellung des Signals für die Erzeuger. Die Gemeinschaft bemüht sich darum, daß man sich der Realität des Milchmarktes wieder bewußt wird. Zum gleichen Bewußtwerden will man auch die Erzeuger von Rindfleisch nötigen, damit sie ebenso wie die Erzeuger von Geflügel und von Schweinefleisch sowie die Landwirte auf die Marktbedingungen achten.

Das bedeutet nicht, daß die Erzeuger auf den Markt losgelassen und sich selbst überlassen werden. Die Preise werden weiterhin gestützt und der Steuerzahler wird diese Last weiter tragen müssen, aber nun nicht mehr für eine unbegrenzte Menge von Erzeugnissen. So wird die gemeinschaftliche

Agrarpolitik den heutigen wirtschaftlichen Gegebenheiten angepasst und der Agrarsektor kann zur Umstrukturierung der Wirtschaft in der Gemeinschaft einen Beitrag leisten.

<u>THE INDUSTRIALISATION OF ANIMAL PRODUCTION</u>

Giovanni BALLARINI,

Director of the Clinica Medica Veterinaria,

University of Parma, I-43100 Parma (Italy),

Member of the Scientific Committee for Animal Nutrition

of the European Communities.

Introduction

For about two and a half million years man has been present on Earth.
Animal <u>domestication</u> only started about ten thousand years ago, while industrial farming has only developed over the last thirty to forty years. If one
compares the period of man's presence on Earth to a calendar year, animal
taming has taken place on the morning of the thirtieth of December and farming
industrialisation started a quarter of an hour before midnight on the thirty-
first of December.

We do not know why man only "invented" the domestic animal after two
and a half million years. It can be assumed that man started to tame animals
at almost the same as vegetables, although the event occurred in different
geographical areas and involved various animal species. Afterwards, within
several millenia, he spread over almost the whole surface of Earth. At the
same time, he started to become sedentary, invented the town and began the
process of urbanisation. Demographic expansion began at the same pace and
with ups and downs led to today's situation and the perspective of exceeding
ten billion souls in the forthcoming decades. In ten thousand years, we have
multiplied by about ten thousand.

We know better the steps of the development of industrialisation in
animal husbandry, the first symptoms of which began to appear at the end of
the last century and which flourished after the second world war. As one will
see later, the reasons and consequences of animal industrialisation remind us
of those of animal domestication : they are closely linked to the human social
structure, to human knowledge of techniques, particularly in the field of
agriculture, to urbanisation and the demographic push. Even if industrialisation has more consequences than taming, there is a continuity between these
two phenomena which has been much too often neglected particularly for the
following reasons :

a) the speed with which industrialisation developed and replaced traditional farming systems;

b) the difficulties encountered by individuals and society in general in adapting mentalities and habits to new situations, particularly when changes are faster than one generation;

c) the development of town population which only knows urban animals and ignores more or less anything related to agricultural or animal production on which it depends for its food.

This paper proposes to illustrate some aspects of animal farming industrialisation with some personal considerations.

Intensive animal farming and technological animals

The industrialisation of animal production has led to the creation of intensive or industrial farms.

Intensive farms are identified by a series of characteristics which usually appear simultaneously. Following are the most important ones :

1.   Very high specialisation of animal species and production targets;

2.   Use of animals from highly selected breeds and varieties in relation to the type and quality of production required and to breeding technologies (technological animals);

3.   Division of animals into buildings particularly equipped for each breeding operation;

4.   Large use of technologies for the control of the environment, the distribution of feeds, the removal and treatment of manure, the harvest of production, etc., with the aid of mechanical devices, nowadays monitored by computers;

5.   Use of specialised feeds according to age or production times, established by advanced computerized mathematical methods, prepared by highly automated mechanical methods, using numerous straight feedingstuffs and industrial by-products, largely exploiting the inter-supplementation between the various feeds and the activity of feed additives;

6.   Use of outside sources of energy for heating, ventilation, tools, etc.;

7.   Need of capital for the buildings and cost of exploitation, which will be higher than for traditional operations;

8.   Use of specialised personnel which is very limited in number proportionally to the numbers of animals on the farm and working for the same number of hours as workers in industry;

9.   Careful control of infections, infestations and pathology;

10.   Efficacious accountancy and technico-economical administration for

each production stage and total production.

Intensive livestock farming achieves high productivity compared to the invested capital and manpower. It also enables a better exploitation of available acres and feeds which are not intended for human usage and which could not easily be used for animals in traditional farming. Furthermore it responds to the demands for food of animal origin in large quantities and of high quality. To understand better why and how intensive units developed in all industrialised countries (independently from any socio-political consideration), some aspects of the problems must be analysed.

Intensive livestock farming permits us to breed animals with limited manpower. For example, if in 1950-1953 it required about 40 hours of work per year to produce 100 kg of pig ready for slaughter, in 1975-1980 it only required four hours and now with increased mechanisation and computerisation the time required may be reduced to two hours. Improvement of productivity is ten to twenty times greater. The exploitation of the benefits of this improved productivity has permitted the reduction at the same time in the number of workers, the number of working hours of the remaining personnel and the increase in their wages. Work conditions and monthly wages of farm workers now tend to be level with those of workers in the non agricultural industry. From a general point of view, through increased productivity, the industrialisation of breeding has permitted altogether an increase of total animal production and a sensitive reduction of manpower, which is typical of all the most developed societies. The problems that this poses are those of the equilibrium between urbanisation and rural exodus - the general and farming industrialisation qualification and permanent education of farm personnel.

Intensive stock farming is characterized by a heavy concentration of animals. It also enables us to exploit optimally feeds produced by intensive agricultural systems and industrial by-products which may come from far away. This type of breeding is thus also possible when land availability is limited. In the European Communities, the average cultivated surface per inhabitant is approximately 0.40 hectares, ranging from 0.16, 0.19, 0.22 in the Netherlands, Belgium and Luxembourg, and the Federal Republic of Germany to 1.04 and 1.99 in Greece and Ireland respectively (Table 1). Intensive breeding is the only effective response to a concentrated human population living on a territory with a relatively limited quantity of arable land which is further eroded because of the occupation of the area by industry, roads and houses.

The production of food of animal origin has always posed problems of food competition between man and animals. In the past animals which used feeds which man could not use were privileged : grass from pastures for ruminants, kitchen wastes and odds for pigs, uncultivable land for semi-wild birds

TABLE 1

Percentage of agricultural population compared to the total working population
and arable land per person in the Member States of the European Communities
(1980)

STATE	Agricultural population % of total working population	Arable land per person (ha)
Belgium	3.0	0.19
Luxembourg	6.6	0.19
Denmark	8.3	0.60
France	8.8	0.63
W. Germany	8.0	0.22
Greece	30.3	1.04
Ireland	19.2	1.99
Italy	14.2	0.32
Netherlands	4.6	0.16
United Kingdom	2.6	6.33

and game. Animal domestication accompanied agriculture in the improvement of
vegetable availability. One thus managed to feed animals with nutrients which
could feed man. At present, for economical reasons also intensive animal
production makes use of feeds which are improper for human consumption for
physiological reasons (grass eaten by ruminants) or psycho-sociological
reasons (distillery by-products which can feed omnivorous animals).

In intensive livestock farming, as it is possible to control almost
all the production factors, the genetic factors of animals, and thus the "new"
breeds developed by man for this purpose, are largely exploited (technological
animals). The improvement of production performance in all stock farming
sectors is formidable. For example, in 1930 a pig reached the weight of 65 kg
after 140 days, while now in the same period it reaches 140 kg, an improvement
of 215 %. At the same time, the quantity of feeds required has decreased,
mainly in products which could be used by man. High production increases have
been achieved in milk from dairy cows (Table 2), in broiler chickens, fish,
etc. As a consequence, the income of livestock farmers has increased and
also, more important, the availability of food of animal origin for the human
population, at progressively lower prices compared to the average price index,
with a progressive decrease of the incidence of the food bill on the whole
family budget. As a result there is a larger quantity of food of animal
origin at lower prices.

At the same time, the possibility of controlling production factors
(genetic, environment, feeding, health, technology, etc.) has permitted us to
address on a more rational basis the problem of the quality of human food.

TABLE 2

Increase of milk production. Annual production per cow in the EC Countries (Kg/head) (Source : ASSOLATTE).

STATE	1970	1980	Variance % 1970-80
Belgium and Luxembourg	7,092	7,818	+ 10.2
Denmark	3,490	4,846	+ 23.0
France	3,110	3,720	+ 19.6
W. Germany	3,800	4,552	+ 19.8
Greece	–	–	–
Ireland	2,504	3,234	+ 29.2
Italy	2,642	3,362	+ 27.3
The Netherlands	4,340	5,026	+ 15.8
United Kingdom	3,891	4,757	+ 22.3

Nutritional, sanitary, technologic, gastronomic quality and acceptability for the consumer may be controlled and improved by current techniques in intensive stock farming (Tables 3 and 4). For example, it is possible to modify the quantity and quality of fats in meat, their colour, their resistance to rancidity, and so on. Finally a high degree of standardisation of production can only be achieved in intensive animal production.

Only one type of animal is bred : the technological animal which has been rigorously selected for a certain type of production. This selection is made in the intensive farms where conditions are adapted to animals which have been selected to fit in with the environment. For example, when developing a milking machine which simulates the calf's sucking, one had to select cows with teats that would adapt to the uniformity of the milking machine.

TABLE 3

Improvement of animal production performance and quality.

Animal species	Meat, milk and eggs with 100 kg of animal feeds	
	1960	1978
Chickens	20 kg	50 kg
Turkeys	15 kg	28 kg
Pigs	23 kg	32 kg
Cattle - milk	8 l	16 l
Layers - eggs*	120 eggs	220 eggs

* The feed consumption per egg has decreased from 275 g to 150 g.

Source : Valfrè, F., 1982.

TABLE 4

Production benefits following the use of various antibiotics (average values in research conducted in recent years)

Animal species	Production parameters	Total number of experiments	Total number of animals	Average improvement in treated animals
Calves	Weight gain	11	400	8.6 $\pm$ 1.4
	Feed efficiency	10	350	6.0 $\pm$ 0.9
Pigs	Weight gain	197	48,000	6.5 $\pm$ 0.5
	Feed efficiency	195	48,000	4.1 $\pm$ 0.4
Broilers	Weight gain	138	200,000	5.0 $\pm$ 0.6
	Feed efficiency	138	200,000	3.5 $\pm$ 0.4
Turkeys	Weight gain	25	30,000	6.3 $\pm$ 0.8
	Feed efficiency	25	30,000	3.0 $\pm$ 0.6

Source : Valfrè, F., 1982

An important consequence of all this is the "technological animal - intensive breeding system". This "system" has its obvious autonomy, particularly where animal welfare is concerned. As well as the welfare of a citizen in a developed country is considered in a different manner from that of the hunter-fruit picker of a primitive population, it must be considered that the welfare of a technological animal that lives in an intensive unit is different from that of the domestic animal on a traditional farm or of the wild animal in its natural environment. Any transposition, and above all any anthropo-morphic elaboration, must be considered as erroneous and a source of equivoca-lity.

Industrialisation and technopathies

The relationships between industrialisation and pathology are very close. Firstly industrialisation and setting up of intensive units have only been possible once a very efficacious control existed against the "important diseases" which used to be able to annihilate traditional stock farming. Let us for example remember anthrax, cattle foot-and-mouth disease, swine pest, chicken pest and coccidiosis. The same phenomenon occurred with man for whom the development of urbanisation (town = human intensive breeding) progressed at the same pace as the control of large epidemics (black death, cholera, typhus, smallpox, etc.). In intensive stock farms, disease is considerably reduced compared to its incidence on traditional farms or in a natural environment. Reduced levels of disease are indispensible on intensive farms to achieve a good level of productivity.

However, disease has not completely disappeared. Actually, new forms of disease have appeared : technopathies. As the word says these are diseases (pathies) which are conditioned by breeding techniques (techno). In this context, some considerations on the relationship between techniques and diseases become necessary.

The main target of evolution is to achieve a "harmony" between the living beings and the environment and also amongst living beings. Therefore, the organism has elaborated complex systems. The immune system regulates and harmonizes the relatioships between animals and parasites, bacteria or viruses. The hormonal system has elaborated means of adapting to the environment. This is so for all organic systems. However, the relationships are not always perfect and the adaptation to the milieu is always incomplete due also to continuous changes in its various components. In natural situations, disease is a means of eliminating the less adapted animal. Many viral diseases, for example rabies, control animal populations and prevent overpopulation. In sufficiently standardized situations, when man can control the concentration of animals, such as in intensive units where technological animals are bred, the adaptation is undoubtedly high, but never perfect. The risk of disease thus persists. In addition, in the "technological animal – intensive breeding system" that we have just described, new conditions have been introduced which were of limited significance in the past. Following are the most important ones :

1.   A reduction of individual variances. The group of technological animals is constituted of animals with similar characteristics and which respond to the environmental variations according to the "all or nothing" model. It is thus sufficient to bring relatively small variations to the milieu in its broad sense to see disease occurring in almost all animals.
2.   An increase of some nutritional requirements. The rapid rhythm of growth and high levels production (eggs or milk) is the result of metabolic changes. One will thus not be surprised to see superior or different nutritional requirements of technological animals from those of traditional or wild animals. The example of vitamin requirements is typical.
3.   A reduction of the adaptability. Metabolic overwork resulting from high production performance reduced somewhat the adaptability of the technological animals. All this is obviously a consequence of the hormonal response in laying hens or dairy cows with high performance.
4.   A decrease of organic defence capacities. A selection mainly aimed at animal production in increasingly "abiotic" environment (through the reduction of infectious and parasitic diseases) and the intervention of man who saves

large numbers of animals from death through the use of medicines, particularly
when their genetic value is high, have reduced the selective pressure which
aimed at producing animals capable of a strong and rapid immune response.
Technological animals have consequently a reduced natural immunity, which is
even more dangerous due to the presence of immunodepressors.

5.   A reduction of contacts between animals of different species which
makes the transmission of similar diseases impossible from an immunity point
of view. In nature, for example, the living in common of bovines, dogs and
humans permitted the circulation of three immunologically similar viruses :
bovine pest, young age disease and measles. This was also the case for
smallpox. In this manner, the natural environment performed an exchange of
apathogenic virus amongst non specific species which developed a group
immunity. All this fails in intensive units where only one animal species is
present.

6.   Movement of animals on a worldwide scale. These movements spread
infections which in the past were limited to particular areas. This is
typically the case of transmissible infectious bovine leucosis.

     Technopathies result from the abovementioned conditions. These are
"old" diseases which have changed their pattern of diffusion, their severity
and the type of clinical symptoms in the new situations. These can also be
"new" diseases caused by bacteria or viruses which the animal used to stand
quite well under traditional farming conditions and which were subclinical
diseases (opportunistic infections). Under conditions of intensive farming,
mastitis for instance belongs to the former and respiratory conditions belong
to the latter. One cannot forget that many technopathies are taken into
consideration because their importance under conditions of intensive farming
has been recognized. There is no doubt, for example, that milk production
problems have always existed, but they were negligible when the cow only
produced a few litres of milk. Nowadays, they have a significant importance,
because milk production has increased and the "quality" of the milk that is
bought from the farmer has acquired a new importance.

     Technopathies do not affect the farm to the same extent as the
"important diseases" did earlier. They can nevertheless diminish more or less
strongly the productivity of the intensive unit. Indeed they decrease the
quantity and the quality of the production. This is why they are also called
"economic diseases". From this point of view technopathies must be considered
as an obstacle to the development of intensive stock farming and their control
is a most important factor for the productivity of the business.

     Traditional veterinary medicine cannot diagnose completely and satis-
factorily the technopathies. Indeed at first examination animals appear to

be healthy and production seems normal. On the contrary, if one measures the quantities produced, if the quality is analysed and if other specific examinations are performed, one can observe technopathies. For example, subfertility on a farm is a technopathy : the animals reproduce, but a careful examination will show that reproduction does not follow the desired rhythm. If a cow does not give birth to one calf each year, but only every 13 to 14 months, there is a 8-16 % infertility; if a sow does not produce twenty piglets per year, but only sixteen, there is a 20 % infertility. The same can be said for the other animal species and the other forms of production (meat, milk, eggs, wool).

In addition, veterinary medicine cannot control technopathies with traditional tools. To control a technopathy means to intervene when the damage has already been observed. Drugs even can only be used to limit the damage. It is thus necessary to establish systems of continuous monitoring to see early the symptoms of a technopathy. It is particularly important to have preventive systems.

Technopathies reveal that the technological animal has not adapted perfectly to intensive stock farming. The most important means to harmonize the technological animal and the intensive farming system are as follows : a) professional qualification of the farm manager; b) automation of the breeding operations and, more important, their accurate functioning which is only possible with electronic computers; c) the use of molecules which can protect the organic activities and which can harmonize the organism on one side and the micro-organisms and parasites on the other; d) lastly the use of biological immunity systems. The incidence of technopathies considered as the signal of an insufficient or of the absence of harmonization constitute a very important element to reveal the incomplete efficiency of intensive breeding either from qualitative or from the quantitative point of view.

From this standpoint the disease has the same meaning in intensive stock farming as in natural (or traditional farming) conditions : in both situations the disease reveals a lack of harmony which negatively affects the productivity of the system.

Hygiene and drugs

It has already been mentioned that industrialisation of farming has become possible once the "major diseases" could be controlled and that the productivity of intensive farming is not possible without an efficacious control of technopathies. Only rigorous hygiene will enable the achievement and maintenance of these objectives. Animal production is thus only applied hygiene.

The most important hygienic conditions in intensive stock farming are as follows :

1.   Specialized and appropriate breeding installations which correspond to the physiological and production requirements of the animals (building hygiene : ventilation, disposal of manure, etc.).

2.   Division of the operation according to the various production phases to achieve what is said above.

3.   Division of animals into homogenous groups of a limited number of animals to avoid overpopulation.

4.   Breeding animals in "separate lots".

5.   Systematic "all in-all out" application with systematic disinfection of the buildings.

6.   Maintenance of the animals in the best possible conditions of feeding and nutrition to achieve an optimal immune response.

7.   Systematic usage of treatments against endo- and ectoparasites.

8.   Large use of preventive treatments such as vaccines or chemoprophylactic drugs (coccidiostats, etc.).

In connection with breeding hygiene as described above, the use of drugs must also be discussed both from a therapeutic and an hygienic-prophylactic point of view.

The therapeutic use of drugs in intensive farming has already been mentioned in relationship with treating an animal of value (individual therapy). Their therapeutic use applied to a group of animals or to a sector of the stock farming will help limit the damage caused by disease. It cannot be forgotten that without their usage a whole farm could have been almost totally or totally destroyed as was the case before the introduction of antibiotics. Fire can be a good example : in modern buildings, particularly the large ones, preventive measures against fire are already incorporated in the plans in order to detect and limit fire rapidly (preventive hygiene in intensive stock farming serve the same purpose), but at the same time fire extinguishers continue to be foreseen and fire brigades to be trained (veterinary drugs are kept to be used for the same reason). The possibility to have drugs available to face emergency cases is thus absolutely necessary.

In intensive stock farming, drugs play a major role as hygienic-prophylactic means against technopathies. Technopathies must be prevented by technological means involving genetics, environment, management, nutrition, but biological and biochemical drugs must also be used. The importance of the drugs increases when they become an element of breeding technology. Everyone now accepts that disinfectants and external antiparasitic drugs are technological elements. Similarly, internal antiparasitics, almost all vitamins and feed additives, as well as the drugs which have an activity on the gut or ruminal microflora and thereby improve feed efficiency, must be considered part of intensive farming technology. The latter drugs (so-called auxinical, growth promoters or permitters, performance promoters) while they sometimes have an antibiotic activity are not therapeutic due to their spectrum of activity and the dosages that are used. Presently one can also consider as technological elements of intensive farming the hormones and related products, as well as castration used to be a production tool in traditional farming. The use of "technologic" type drugs is indispensible in intensive farming and is an integral part of it. How would it be possible to breed one hundred thousand chickens without the use of a coccidiostat ?

In intensive farming, the choice of pharmacological treatments, their possible association and above all the use conditions (timing and duration of the treatments, etc.) must take various aspects into account. The following are the most interesting ones :

1.   The type of farming and the technologies used.

2.   The coordination with other possible hygiene measures.

3.   The relationship between the cost of intervention and the benefit that can be drawn from it.

4.   The safeguard of production quality or quantity.

5.   The absence of risk for the consumer (problem of residues and metabolites), for the farm worker and also for the environment.

Precise national and international conventions define the safety limits for the choice and use of drugs, taking into account the type of usage : individual or group therapy; technological usage.

Some considerations must be made about health. The first is that the good result of drug usage in intensive stock farming is often conditioned by the farm hygiene and the use conditions. All this confirms the importance of the role of "technological" drugs in intensive farming. These will thus have to be considered as indispensable elements as well as the structure of the farm environment.

The motto : "More hygiene, less drugs" has often been cited. This is of course a fascinating motto and we should not lose sight of it. Nowadays it is mostly considered as a good means of avoiding therapeutic measures, but one cannot think that this objective is about to materialise. Let us come back to our example of the fire : in spite of all the precautions taken at the time of building, extinguishers and emergency exits and ladders will have to be kept.

More hygiene, less drugs does not apply to what we called technological drugs : vitamins, feed additives, modulators of the microbial flora of the guts or rumen, hormones and hormonomimetics, etc. which are used on "healthy animals". These drugs will have to be considered as essential parts of the structure of intensive stock farming. One could not eliminate them unless one changed the type of farming, but to date the experiments of "specific pathogen free" or "germ-free" animal farming have not yet found a practical implementation on a large scale. On the other hand, in the "minimal disease" type farms, the technological drugs are still perfectly valid and justified.

A last consideration is that of the risks that the use of drugs in animals may cause to man or the environment. There is no doubt that the correct usage, with the indispensable minimum dosages and the shortest duration of treatment is easy in intensive units. In addition the public authorities can perform efficacious controls on these farms. The real risk comes from the smaller more or less "traditional" farms where for many reasons abuse is easier, more frequent and widespread.

Conclusion

According to data of the European Communities in 1960, the Community of the Six counted 15.2 million persons involved in agriculture and animal production, while in 1973 there were only 8.2 million. This means that over that period of time the farm worker population has decreased practically by one half. As an average, from 1960 to 1973, every minute one peasant left the country. Many of these workers have been absorbed by other fast developing economic sectors (industry, services) which needed additional manpower. From 1973 to 1980, after the enlargement of the Community, the agricultural population has further diminished from 9.9 to 7.4 million, i.e. by 7.4 % of the total working population, at a reduced pace compared to the beginning which in the last decade corresponds to 2.8 % per annum. This slowdown is not surprising if one considers the general economic crisis and the level of unemployment. In the present situation, the decrease of the agricultural population can be accounted for by the death or retirement of heirless farmers rather than by rural exodus of active persons to other sectors.

In spite of the large numerical concentration of persons occupied in the last two decades, the economic efficiency of the agricultural and animal production has increased nominally by over 7 % per annum. This increase is inferior to that of general economy achieved over the same period (+ 10.5 %). It shows however a spectacular expansion of work productivity in European agriculture and animal production compared to the strong decrease of personnel.

In spite of the fact that the needs of the Community, particularly where animal products are concerned, have continued to increase (table 5)

TABLE 5

Food consumption per person in the Community Countries (kg/year)

COUNTRY	1970-71(°)			1978(°°)			Variance % 1970/71-1978		
	Meat	Milk	Eggs	Meat	Milk	Eggs	Meat	Milk	Eggs
Belgium and Luxembourg	79	80	15	95	74	14	120	92	93
Denmark	62	123	11	80	140	12	129	114	109
France	95	70	15	108	79	14	114	113	93
W. Germany	85	79	16	97	74	17	114	94	106
Greece	–	–	–	(68)	(100)	(13)	–	–	–
Ireland	84	213	13	92	204	11	110	96	85
Italy (°°°)	50	66	11	72	79	12	144	120	109
Netherlands	59	107	12	76	112	11	129	105	92
United Kingdom	73	142	15	74	139	15	101	98	100
Average	73.4	110.0	13.5	86.7 (°°°°)	112.6 (°°°°)	13.2 (°°°°)	118.2	102.4	98.1

(°) Annuaria Statistico Italiano
(°°) Euroforum
(°°°) Some data differ from national statistics.
(°°°°) Greece excluded.

in many sectors the self-sufficiency quota has been maintained and even
exceeded. The increase in needs has been favoured by the decrease of
agricultural and animal production prices to producers and consumers compared
to the general price index. Indeed, the prices of agricultural and animal
production have increased at a much slower pace in the Community than the
individual revenue. As can be seen in table 6, to a 12 % increase of
individual revenue in the period 1973-1980 corresponds an increase of food
prices to consumer of only 10.8 % and to producers of 7.4 %. One can thus
rightly talk of a supply to consumers at reasonable prices.

TABLE 6
Average yearly price increases in the European Communities from 1973 to 1980

Prices	Average yearly increase %
Agricultural and animal productions - price to producer	7.4
Food price to consumer	10.8
General price index	11.2
Net revenue per inhabitant	12.0

Source : EC, 1982.

These results are no miracle, but are for a large part a consequence of the increase of productivity linked with specialisation and intensification. In the field of production of food of animal origin, specialisation and intensification are highest in the intensive farm units.

It must further be stressed that the basic indication of a recent report of the European Communities is that the number of persons employed in agricultural and animal production is going to further decrease if productivity must increase. Increasing productivity is still necessary in spite of the already notable improvement performed in the agricultural and animal production. In the last twenty years, productivity in general has improved by less than 4 % while in the agricultural sector it has exceeded 6 % per annum. In addition to have a revenue of 1,000 ECU in 1970 one had to employ 220 workers, while in 1980 this can be achieved by 150 persons. The differences between the various geographical areas remain nevertheless important and this partly explains the increasing differences and different production prices in the various areas.

Industrialisation of stock farming is today the only concrete response to the demands of the consumers and producers. Consumers require abundant supplies of food of animal origin, of high quality and at reasonable prices. Producers need to make profit. Increasing productivity in intensive farming is still possible, but it requires more interdependence between the technical and economical sectors upstream and downstream, as well as an improved preparation of the workers.

As specialisation and intensification increase, so does the interdependence with industry for investments (buildings, installations, tools, etc.) and production tools (fuel, electric energy, maintenance and repair of the installations, feedstuffs for the animals, additives, antiparasitics, disinfectants and drugs). In this context it is necessary to stress once more that drugs, particularly those that are used systematically (the above mentioned technological drugs) must be considered as actual production tools.

The relationships with various sectors of farming downstream become very important. The agricultural and animal products introduced on the market in their natural state are less and less frequent, as they are for a large part transformed. In the marketing and transformation phases, the quality and its specifications are controlled. As already said such controls are easier with intensive production. It must be emphasized that only intensive production permits one to respond rapidly and exhaustively to the quality demands of the transformation industry and the consumer. Indeed it is possible today to product standardised food of animal origin which have all required characteristics : fat and lean meat, fats with particular nutritional or technological qualities, milk with certain characteristics for industrial usage, etc. In all cases there is a higher sanitary quality. In general, it only depends on the intensive producer to exploit fully his capacity to produce the required quality. Apart from some well defined sectors, such as milk destined for cheese production, consumers are not always able to formulate clearly, univocally and precisely their quality requirements. In general they express a vague wish for "honesty" and "natural products", which lack precision and can induce equivocality by their lack of scientific definition. On the contrary, intensive farming is able to produce food of certain quality and in the required quantity, but one may not forget that quality has a price, and this price must be paid.

Productivity increase in intensive farming, including the production of high quality food, is not possible if the professional qualification of farmers is not improved. It is indeed necessary that farm workers should be able to exploit the numerous possibilities offered by research and techniques. A permanent education of farmers is thus necessary, as well for technical as for economical or health problems.

In this area, new responsibilities may face the farmer. Let us consider for example the situation of drug usage : it is increasingly obvious that the farmer is responsible directly and definitely for the choice of therapeutic and above all technological drugs. There is no doubt that researchers, producers and all those who authorize the usage of drugs accept this responsibility : farmers must do the same. All this to protect human health and the natural equilibrium of the environment.

Agriculture and animal farming are the oldest activities of "modern" man, i.e. of man in the last ten millenia. Their products are destined for human food and they have a vital importance for the "quality" of life. Productivity improvements are an essential prerequisite for the development of other sectors of activity. It is only if those involved in the agricultural and animal production sectors produce more than for their own needs that other men can devote themselves to other tasks than food production. Industrial development separate from agricultural development is thus impossible. At the beginning agricultural development permitted the movement of part of the population to other sectors, but later on it is the industrial sector which permits a further development (industrialiation) of agricultural activities. In this perspective, intensive farming is strategically important.

One cannot conclude this presentation without talking of "post industrial evolution" of animal production. My personal opinion is that technical conditions already exist for such an evolution, but there is still much to be done in the sociological and cultural areas. It is possible indeed to envisage a certain "saving" of proteins of animal origin, but beefsteak is a symbol of wealth which nobody wants to forsake. It will also be possible, in the near future, to use proteins of fermentation origin directly in human food, but the violent visceral response of many countries proves that we are a long way away from this "post traditional" nutrition.

Man has been carnivorous for some two and a half million years. It will be very difficult for him to change this characteristic which grows its roots in its very nature. A large production of meat and other food products of animal origin is thus indispensable to respond to an increasing world population. Only industrial animal farming will be able to respond to their demand for good quality food in a more or less short delay (ten, one hundred or one thousand years ?).

References

E.C. Report, 1982. The Agricultural Policy in the European Community, 6.
Valfrè, F., 1982. Les vues des utilisateurs d'additifs. Conf. Vétérinaire
    "L'industrie Pharmaceutique et la Santé Animale", Rome, 20-22 octobre.

RESUME

L'INDUSTRIALISATION DE LA PRODUTION ANIMALE

Giovanni BALLARINI,
Directeur de la Clinica Medica Veterinaria,
Université de Parma, Italie,
Membre du Comité Scientifique de l'Alimentation Animale
des Communautés Européennes.

L'industrialisation de la production animale a permis de produire de plus grandes quantités d'aliments d'origine animale, de qualité supérieure et à des prix relativement bas. Dans le même temps, il s'est opéré en parallèle une diminution considérable de la population agricole et une augmentation de la productivité des élevages. Ce phénomène est intimement lié à la structure de notre société industrielle et a amené des modifications sur plusieurs plans : social, économique, technologique, écologique et sanitaire.

L'existence de l'élevage industriel et intensif est subordonnée à un contrôle efficace des pathologies. Sans ce contrôle apparaissent non seulement les maladies traditionnelles, mais également des affections conditionnées par la technologie de l'élevage (les technopathies conditionnées). Apparentes ou subcliniques, les maladies entraînent, pour le producteur, des pertes dues tant à la réduction de la quantité qu'à une diminution de la qualité des produits destinés à la consommation humaine.

Le contrôle des technopathies est basé, d'une part, sur des méthodes technologiques (génétique, gestion, hygiène alimentaire et hygiène de l'environnement) et, d'autre part, sur des méthodes biologiques et biochimiques.

L'utilisation de produits biologiques en prophylaxie ou en thérapeutique a pour objectif de prévenir ou de réduire les dommages causés par les maladies cliniques ou subcliniques. Tous les moyens de contrôles sont sélectionnés en fonction du coût des interventions et des bénéfices résultant de la plus grande production de denrées alimentaires ou de leur meilleure qualité suite au traitement des animaux. Des réglementations nationales et internationales précises établissent les limites de sécurité des traitements pharmacologiques.

Seul l'élevage industriel permet de produire une nourriture d'origine animale à un prix raisonnable et surtout à des niveaux de qualité prédéterminés : valeur nutritive, qualité sanitaire, qualité technologique, qualité gastronomique, etc. Cependant, la qualité doit se payer. D'autre part, si la demande s'en fait sentir, l'éleveur peut encore l'améliorer par des interventions de type génétique, diététique, gestionnel, etc. Il est, toutefois, indispensable de définir les niveaux de qualité de façon précise.

Un très bon état sanitaire des animaux est également important dans les élevages industriels, tant du point de vue hygiénique qu'écologique, et plusieurs molécules (comme les vitamines synthétiques, par exemple) sont devenues des moyens technologiques nécessaires et ne sont plus utilisées comme des médicaments.

La complexité de l'élevage industriel exige enfin des qualités professionnelles très élevées de la part des producteurs dont la responsabilité vis-à-vis de la santé publique est toujours plus grande. C'est pourquoi il est nécessaire d'assurer l'éducation permanente des éleveurs.

ZUSAMMENFASSUNG

DIE TECHNISCHE VORAUSSETZUNGEN DER TIERISCHEN PRODUKTION

Giovanni Ballarini
Direktor der Clinica Medica Veterinaria
der Universität von Parma, Italien

Die Industrialisierung der tierischen Produktion hat es uns erlaubt, tierische Nahrungsmittel in größeren Mengen, in höherer Qualität und zu verhältnismäßig geringen Preisen zu produzieren. Gleichzeitig hat die Zahl der in der Landwirtschaft beschäftigten Menschen abgenommen und die Produktivität der Zuchtbetriebe zugenommen. Diese Erscheinung hängt sehr eng mit der Struktur unserer industriellen Gesellschaft zusammen und hat zu Veränderung auf verschiedenen Ebenen geführt, z.B. auf der sozialen, wirtschaftlichen, technologischen, ökologischen und sanitären Ebene.

Die Existenz einer Intensivaufzucht im industriellen Maßstab hängt von einer wirksamen Kontrolle der Krankheiten ab. Ohne diese Kontrolle treten nicht nur die herkömmlichen Krankheiten, sondern auch Pathologien auf, die durch die Aufzuchtbedingungen entstehen (Technopathien). Mögen diese Krankheiten nun offen zu Tage treten oder in subklinischer Form vorhanden sein, sie führen beim Produzenten zu Verlusten in quantitativer und auch qualitativer Hinsicht.

Die Kontrolle der Technopathien beruht zum einen Teil auf technologischen Verfahren (Genetik, Haltung, Hygiene in Bezug auf Nahrungsmittel und Umgebung) und auf der anderen Seite auf biologischen und biochemischen Verfahren.

Die Verwendung biologischer Produkte bei der Prophylaxe oder der Therapie hat zum Ziel, die durch klinische oder subklinische Krankheiten verursachten Schäden von Anfang an zu verhindern oder zu vermindern. Alle Kontrollmittel werden im Hinblick auf ihre Kosten und ihre Auswirkungen ausgewählt, wobei die Behandlung der Tiere zur größeren Quantitäten und zu höherer Qualität der Nahrungsmittel führen sollte. Genaue nationale und internationale Verordnungen legen die Toleranzgrenzen dieser pharmakologischen Behandlungen fest.

Nur eine Aufzucht auf industrieller Ebene kann zu einem vernünftigen Preis genügend tierische Nahrung zu einem im voraus bestimmten Qualitätsniveau produzieren : Nähwert, hygienische und technologische Qualität, gastronomischer Wert usw. Qualität hat allerdings ihren Preis, und den muß der Verbraucher dem Produzenten bezahlen. Sollte sich auch eine Nachfrage danach einstellen, so kann der Produzent durch Maßnahmen auf dem Gebiet der Genetik, der Fütterung, der Haltung usw. die Qualität weiter erhöhen. Es ist dennoch notwendig, die Qualitätsniveaus genau zu definieren.

Ein sehr guter Gesundheitszustand der Tiere ist in industriellen Aufzuchtbetrieben gleichermaßen wichtig, sei es vom hygienischen wie vom ökologischen Standpunkt aus, und zahlreiche Stoffe, z.B. synthetische Vitamine, sind zu unabdingbaren technologischen Hilfsmitteln geworden und werden nicht mehr als Medikamente gebraucht.

Die Komplexität der industriellen Aufzucht erfordert schließlich einen sehr hohen beruflichen Stand der Produzenten, deren Verantwortung gegenüber der öffentlichen Gesundheit immer mehr zunimmt. Aus diesem Grunde ist eine dauernde Schulung der Züchter unerläßlich.

THE QUALITY OF FOOD AND HUMAN NUTRITION

Marcel VANBELLE,

Faculté des Sciences Agronomiques - U.C.L.

Laboratoire de Biochimie de la Nutrition

B-1348 Louvain-La-Neuve, Belgium

Introduction

The concept of food quality is nowadays one of the main preoccupations of numerous consumers' associations. Man has always had to take food to remain alive, to ensure the essential physiological functions, but the evolution of animal and vegetable production and the improvement of the standard of living have resulted in a significant increase in the demand for a variety in the choice of foods. Productivity has considerably improved since the second world war as a result of the formidable progress in the various agricultural sectors. All this has led to an evolution in the meaning of the word "quality" itself. Let us briefly review this evolution.

Evolution in nutrition and nutritional habits

In the past, the consumer used to give more importance to the pleasant sensations brought by food. The gustatory and gastronomic qualities were essential. Contemporaneous of Voltaire, Jean-Jacques Rousseau, Lavoisier and La Place, the famous connoisseur, Brillat-Savarin, wrote in his book "Physiologie du Goût" (Physiology of Taste) : "Gastronomy is the reasoned knowledge of all that relates to man's food" and also "Gastronomy belongs to natural history, by its classification of nutritive substances, to physics by the examination of their composition and quality, to chemistry by various analyses and changes they undergo; to cookery by the art of preparing foods and making them pleasant to taste; to commerce by the search for means to buy the cheapest goods and to sell most advantageously what it offers; and lastly to political economy by the resources it offers in tax and the possibility of exchanges it establishes amongst nations." "Gastronomy rules the whole life : for the tears of the new-born child call for his nurse's breath and the dying man still receives with some pleasure the last potion that he will unfortunately not digest." This is the summary of all an era !

Must this be considered in the light of today's industrial society ?

Some precisions are required, since industrialisation, the conditions of modern living have resulted in an evolution towards a reduction of energy requirements (Dupin, 1984).

Indeed, in the last decades, economic and social changes, the progressive substitution of physical and muscular activity by mechanical and motorized tools in almost all sectors are slowly amending our total food requirements. The muscular immobilisation hits an increasing fraction of our population, both in town and in the country (movements, professional work). Everything becomes increasingly mechanised and even the housewife does not escape mechanisation. In addition, thermoregulation outlay (struggle against heat or cold) have decreased considerably (houses, workshops, offices, cars, public transport are now well heated). In summary, in the last 30-40 years, energy outlays have decreased sensibly for the whole of the population.

As mentioned, in parallel with this evolution in the way of living, agricultural productivity has never ceased to increase (tractors have replaced animals, knowledge of agriculture and animal production have improved as well as animal nutrition, the application of biological knowledge and genetics to animal and plant selection, chemical and biological fight against parasites, predators, weeds and diseases).

As Fauconneau said recently : "This convergence of progress is firstly due to a methodic programming of technical and scientific research applied to agriculture. But at the same time it results from increased technicality of the agricultural population. It has implied the sliding of part of the labour and brains to the mechanical and automobile industry, to the chemical industry (fertilizers, insecticides, herbicides) and to the animal feedingstuffs industry (stocking, preservation and transformation of products).

However, these increased potentials of agriculture and food industry both from the qualitative and quantitative points of view do permit us, as long as it is wanted, to achieve a better adaptation of human nutrition to the actual requirements.

Changes in conditions of living and work have also induced a deep change in our feeding habits. We did not eat in the same manner in 1940 as we did in 1960 and nutrition in year 2000 will not be the same as in 1980.

If we do indeed eat slightly less calories, the reduction is far less than that of our energy outlays :

1.	The consumption of glucides has decreased considerably in the last decades : the reduction in bread consumption has been more than 50 % in the last forty years and although the rural population still eats more bread that people living in town the difference has shrunk considerably in the last

years. Potato consumption has also decreased in a spectacular manner : 50 %
between 1925 and 1980. This is also the case for dried vegetable. This means
a large reduction of complex glucides (starch and starchy foods with slow
digestion) and a reduction in the consumption of fibrous food. On the
contrary, the consumption of saccharose (fast digestion and absorption) has
increased considerably. It reaches 100 g per day in Belgium and France, and
it is the quantity of invisible sugars incorporated in the food (chocolate,
sweets, jams, cakes, yoghurts, desserts and icecreams) which has increased as
well as of sugars incorporated in drinks (soda and cola drinks contain from 90
to 130 g/litre).

Today it is advised to absorb a quantity of glucides equivalent to
55 % of total calories, but with less than 10 % in the form of saccharose,
since the excessive consumption of fast absorption glucides can have bad
metabolic consequences.

2.   What has the evolution of protein consumption been ?

If the consumption of proteins of vegetable origin has considerably
decreased (as bread or dried vegetables), the consumption of proteins of
animal origin (meats, fish, crustaceans, eggs, milk and derivates) continually
increases. The increase of meat in the total protein consumption has been
spectacular in the last 50 years. In France and Belgium, it has more than
doubled : nowadays consumption in Belgium is about 100 kg/person/year and it
is 6 to 10 kg higher in France ! This is a worldwide phenomenon and together
with FAO we can say that consumption is in proportion with the average income
per inhabitant. However, beef proteins are the most expensive. What is the
origin of this irrepressible attraction for meat, red meat in particular ?

It is both because of the high digestibility and high biolological
value of the protein and the high protein concentration, taste, texture of the
meat are determining factors. Let us add to this the ease of use, the diversification in the available forms, the hygienic quality, its association with
other foods which make of it a first-choice food. This is - and it has not
been emphasized sufficiently - due to the progress made in animal production
in general, in the selection and production of high quality, well balanced
feeds.

Ferrando (1983), the wise philosopher nutritionist, recently said :
"between the tardy, ill-healthy if not cachectic cows, beefs, pigs and
chickens which produced a few litres of milk or a few kilos of poor quality
and tough meat, and rare eggs, and animals nowadays, there is the same
difference as that between the lifetime of people in the Middle Ages (33
years) and today (74 years). There are also all the old fears of hunger and
our certainty to be able to eat well. This is the result of animal production

progress. This should not be overlooked but rather proclaimed continuously."

This evolution is thus beneficial, but one should not exaggerate with the quantities, since some meats bring saturated lipids. It is indeed recommended to include in the diet 12 to 15 % of calories from proteins, one-third of which should come from animal proteins. This figures is much too often exceeded : in France and Belgium animal proteins represent about 70 % of total proteins against 10 to 20 % only in the Third-World.

3.   The problem lies with the evolution of lipid consumption and composition. The consumption is continually increasing and is in the order of 120 to 125 g per day per person in Belgium. In France, they represent up to 42 % of the total calory diet while they should not exceed 30-35 % of the daily calory intake. The problem increases by the much too large proportion of saturated fatty acids compared to monounsaturated and polyunsaturated fatty acids : epidemiological studies indicate that too high an intake of saturated fatty acids significantly increased the risk of ischaemic cardiopathies (myocardial infarction).

In practice, saturated, monounsaturated and polyunsaturated fatty acides should each account for one-third of total lipids. The problem of the excess of arachidonic acid and "trans" fatty acid will be mentioned later. The dangerous issue in our society of abundance is that the place of glucides in the coverage of energy requirements tends to be replaced by lipids with all the risks that this represents, particularly for overweight persons. The same problem exists with alcohol consumption exceeding 80 g per day when 40 g per day would be beneficial (J. Brison, 1982) increasing the level of cholesterol in the HDL fraction (High Density Lipoprotein of the blood), i.e. the fraction which gives protection against infarction, since it transports the peripheric cholesterol to the liver for metabolisation.

Our food habits have been deeply modified in the last decades and the tendency of our society is to exaggerate in some areas (too many calories, too many lipids, not enough crude fibre, etc.) and this poses the problem of requirements and nutritive balance and the dangers of excesses (stoutness, cardio-vascular diseases and cancers of food origin) as well as of restrictive diets (fear to put on weight). Table 1 indicates the recommended energy and protein intake.

However, the bourgeois continues to like the pleasures of food, seeks variation in the composition of menus, loves fine dishes and has developed the culture of "eating well". But he has become more demanding on quality, although we know today several types of feedings and consumers.

TABLE 1

Recommended energy and protein intake (Dupin et al, 1981)

	Kcal/day	KJ/day	Proteins g/day
Male adult			
20 to 39 years of age			
65 kgs.			
8 hours in bed			
8 hours average activity			
4-6 hours sitting			
2 hours walking			
Limited physical activiy	2100	8800	63
Normal physical activity	2700	11300	81
	(3000 FAO)		$\pm$ 20 %
Important activity	3000	12500	90
Very important activity	3500	14600	105
Female adult			
20 to 39 years of age			
55 kg			
Limited physical activity	1800	7500	54
Normal physical activity	2000	8400	60
	(2200 FAO)		$\pm$ 20 %
Heavy physical activity	2200	9200	66
Children and teenagers			
1 to 3 years of age	1360	5700	23-40
4 to 6 years of age	1830	7600	
7 to 9 years of age	2190	9200	55-66 $\pm$ 10 %
Boys : 10-12 years of age	2600	10900	78
Girls : 10-12 years of age	2350	9800	71
Boys : 13-15 years of age	2900	12100	90
Girls : 13-15 years of age	2400	10400	72
Boys : 16-19 years of age	3070	12800	90 $\pm$ 20 %
Girls : 16-19 years of age	2310	9700	70

Types of nutrition and categories of consumers

A large agricultural association has recently issued a booklet on the quality of foods of agricultural origin and mentions 5 types of nutrition and three categories of consumers, the latter chosing one or more types of feeding. (Van de Pitte, 1983).

The types of feeding currently in the fashion are as follows :

1.   Normal or balanced nutrition which approximately satisfies the requirements of classical human nutrition, i.e. that which brings the indispensable foods and respects the balances : this will be discussed later. It is the true dietetic : the correct nutrition to remain healthy.

2.   <u>Conventional and non conventional nutrition</u> amongst which we have partly or exclusively vegetarian diets, spectacular weight control diets, diets based on uncooked foods, fantasy foods, uncoupled foods. A large usage is made of so-called "health products" which are said, sometimes without scientific evidence, that they have exceptional nutritive and curative properties.

3.   <u>Special foods</u> such as dietetic foods which pretend to restore health by a prescribed unpleasant diet.

4.   <u>Biological or organic diets</u> where foods are produced without the aid of any chemical products (pesticides, herbicides) or chemical fertilizers and made without any additives even though these may be authorized by law.

5.   <u>Natural nutrition</u> which is characterized by the consumption of foods which are still in their original state and which are eaten as such or with a minimum of treatments.

The latter two groups represent the "return to nature".

But like Ferrando (1973) we could raise the question : "What is natural food ?"

"The definition is still to be found, as the conditions of vegetation, of harvest, of preservation according to the most ancient traditions do not guarantee this "natural" characteristic in the sense that it is understood now." Does "nutritional naturism" which aims at suppressing all use of chemical agents in animal feeding and preach the return to an idyllic and archaic concept of feeding represent the safe way for man ?

We do not believe it is. Economical, demographic and hygienic reasons impose the use of regulated and controllable modern techniques in animal farming and production.

Categories of consumers

Today's consumers chose one or more type of feeding. In fact in developed countries consumers could be classified into three groups :

1.   The group of those, the largest one, who want a normal, balanced nutrition but who have doubts regarding the "purity" of the products they eat. There is doubt about the quality that is offered to society. What is wanted is food unspoiled by additives or contaminants. This groups exerts pressure on public authority through the associations for the defence of consumers.

2.   The group of those who really want natural products, either by pure and simple conviction, or by philosophy, but most often as a reaction against modern society and its polluting and contaminating industries in general.

3.   There are also the gastronomes who remain faithful to Rabelais and

Brillat-Savarin and whose philosophy is incompatible with the standard industrial food which often lacks gustative quality and who continue to differentiate good butter and margarine, who dislike canned or lyophillized soup and hamburgers, but who like the "crus", farm butter which is so rich in diacetyl, who love steak or T-bone steak when two centimeters of fat must be cut away. They love all pleasures of the table, including great Burgundy wines and crus of Bordeaux that must be served with this food. For them food is synonymous with enjoyment; taking food is a true "libido". Tremolière writes about him : "Nutritional education will convince his mind, but neither his belly nor his heart; he resists alimentary advertising."

We would now like to discuss about the consumers who belong to the category of well convinced people, with a precise philosophy who look for a healthy nutrition. What are the criteria on which they base their choice ? What is important for them to live healthily ? What are the criteria of the consumer in the choice of food ?

1.   A reasonable price, but quality remains most important. Price is not the major factor.

2.   Safe food. "Quality" used to mean tasty or gastronomic. Now the concept represents more the increasing wish to have foods that do not cause any harm to the consumer's health. They must conform to hygiene standards or production standards : absence of additives, absence of contaminants, of noxious residues. Hence some distrust which is not always justified.

3.   For the more educated consumers : equilibrated foods, with a good nutritional value (balance of calories, proteins and amino-acids, lipids, fibres, limited content of salt).

4.   An honest, biologically natural product, easy to use, unpolluted, conform to traditional preparations (bread, yoghurt, wine, beer, cheese) with good gustative quality and bacteriologically harmless.

Needless to say that it is not easy to gather all the above criteria and one must often accept a compromise. This necessitates a continuous and sincere information of the consumer. The competent authorities (Ministry of Agriculture, Ministry of Public Health, professional associations) must take measures to protect the consumer and operate appropriate controls.

Public opinion has become very sensitive with regard to anything that relates to food hygiene. Too frequently the agricultural worker is unfairly accused of using too much pesticide, or using too many additives in an uncontrolled manner or to abuse of hormones.

These complaints are often unfounded, but it remains true that producers must pay attention to this new tendency and to these new demands.

Let us now examine in more detail the quality of food from a scientific point of view.

Definition of food and food quality

a) <u>Definition of food</u> : Prof. Simmonet gave the following definition of food :

"Food is a substance, generally natural and of a complex composition, which, when associated with other foods in appropriate proportions, is capable of ensuring the regular cycle of an individual's living and the persistance of the species to which it belongs."

This definition takes no account of the emotional (sensorial, symbolic, social, psychological and economic) qualities that man imposes in the desire and acceptation of food and which explains why there is such an extraordinary variety in the choice of food all over the world.

One should neither confuse the notion of food which comprises what is being eaten and the notion of nutrient which is the physiological aspect of food.

For Claude Bernard, the "nutrient" is the food substance which is directly and completely assimilable by the digestive tract or the chemical substances which are indispensable for the health and activity of man.

Taking into account the sensorial behaviour of man, Tremolière defines food as "a commodity which contains nutrients, and is thus nutritive, susceptible of satisfying appetite, thus palatable and accepted as food in the society being considered, thus customary."

b) <u>Definition of the quality of food</u> : The notion of quality is even more discussed, more complex and very relative. The definitions vary according to the consumers' assocations, i.e. there are subjective definitions beside the objective definitions which are generally accepted by the nutritionists. As Tremolière said, the commodity must have :
1.   A physiological function, it must feed. This is expressed by its nutritional value (energy, protein, minerals and vitamins).
2.   A psychosensorial organoleptic function, i.e. it must excite the taste, the digestion and the general feelings, it must be desired by man (aroma, colour, texture) and exert an emotive stimulus.
3.   An ethico-intellectual function, i.e. it has a symbolic, social, economic and cultural value. Tremoliere said again some time ago : "A society is built around the way in which it produces and eats foods" and also "The meal is the cornerstone of family and society". Must one go back to the Lord's Supper to understand that eating together, sharing the feast in a brotherly

manner gives all its dimension to our social life, to our soul, and beyond that to our civilisation. The meal and its composition are the expression of our standard of living.

In the light of the modern techniques used, two main demands can be added :

4.   To be free from undesirable substances : dangerous additives, natural or synthetic contaminants.

5.   A reasonable or affordable price.

Let us review the nutritional value of food and the problems of food toxicity.

1.   Nutritional quality, nutritional value : Nutrition teaches us that food must feed man, i.e. bring calories, proteins, the indispensable amino-acids, vitamins, minerals and fibres which are necessary for the maintenance and development of the human body. This is the physiological aspect of food. All this is laid down in nutrition tables and the detailed composition of each food can be found in them.

But our nutrition is composed of several foods, the menu is composed by man himself and theoretically a balanced nutrition is possible, but is rarely applied (except in a medical clinic). Man is not interested in it by "reason" and is more often guided by his taste, which sometimes leads him to make the wrong choice. The variation in the menu, with the aid of education, helps man to eat correctly. However, recent enquiries show that the middle-man of developed countries eats some 10 to 20 % of calories in excess, particularly as lipids. Actually, the saturated or unsaturated nature of the fatty acids is the centre of discussions on civilisation diseases : athero-sclerosis, hypercholesterolaemia, myocardial infarction, etc., and that is not all.

Let us say in addition that the modern techniques of preparation, packaging and preservation of foods have not diminished their nutritional value. On the contrary, as a result of these our nutrition has become more diversified and qualitatively better.

The quality of food depends of course of its chemical composition, but also of the digestive and metabolic availability of the nutrients. These factors may also be affected by alterations that occur during technological treatments and cooking.

Digestibility is the relationship between the quantity absorbed and the quantity ingested.

$$\frac{I - F}{I} \times 100$$

52

It is measured from digestive balances. Depending on whether endogenous
faecal excretion is taken into account or not, one has the actual or the
apparent digestibility.

The digestive utilisation of energy (digestible energy) depends on the
digestibility of the whole of the energetic nutrients : it varies from 95 %
for animal products to less than 75 % for whole cereals in man.

The digestibility of glucides depends essentially on their intestinal
degradation into single sugars by the digestive enzymes (saliva and pancreatic
amylase, lactase, sucrase and maltase of the cells in intestinal brush).

Complex glucides and diholosides must be hydrolysed in glucose, galac-
tose, fructose, which are absorbable by the intestine. The digestibility of
starch varies from 85 to 95 % depending on the origin, the level of extraction
from cereals and the degree of gelatination.

A number of glucides are little assimilated because the enzymes
susceptible of splitting them do practically not exist in normal conditions.
It is the case of oligosaccharides or alpha-galactosides of the seeds of some
Leguminosae (raffinose, stachyose, verbascose). These oligosaccharides are
metabolised by the intestinal flora in the distal areas of the digestive tract
and may be at the origin of certain troubles (aerophagia, ...)

Furthermore, the genetic insufficiency of some disaccharidases leads
to some conditions of digestive intolerance to corresponding disaccharides.

Intolerance to lactose is either due to hereditary deficiency in
intestinal lactase which is observed in certain populations, or to a
disappearance of the enzyme due to age or the lack of milk consumption.

Lastly, cellulose and pectines are indigestible polysaccharides (in
the small intestine) in man because of the lack of specific lytic enzymes.
Chitin of insect or crustacean shells is not significantly digested or
absorbed.

The average digestibility of lipids in man is about 95 % for animal
fats and vegetable oils, 90 % for lipids from cereals and Leguminosae. Oil
hydrogenation (with formation of "trans" acids) lowers the digestive use of
unsaturated acids. The metabolic utilisation of "trans" fatty acids is
slower, their accumulation in the organs is an undisputed fact and there is
some worry about their possible role in atherosclerosis, myocardial infarction
and even carcinogenesis. Now, the presence of "trans" fatty acids in butter
is insignificant (0.1 to 3 %) while in margarine it can be up to 40 %
(Brisson, 1982). Should one not then review the thesis defended in the years
1950-60 that unsaturated fatty acids were favourable to a reduction of hyper-
cholesteremia and myocardial infarction ? The more so that it has recently
been said that an excess of arachidonic acid (C_{20} with four double links) is

at the origin of metabolic production of tromboxane TA_2 and prostaglandins PGF_2 and FGE_2 which favour blood platelet aggregation. Only prostacycline which derives from it too has an antithrombotic activity. But in total the action of arachidonic acid is more negative than positive (Devis, 1981) in the aetiology of atherosclerosis. A matter to be followed up ...

Protein digestibility in man is around 95 to 97 % for meat, eggs, fish and milk. Cereal proteins are 83-85 % digestible, white bread 91 %, Leguminosea is much lower, but varies considerably depending on the vegetable ranging from 60 to 85 %.

However, the nutritional quality of a protein does not depend only on its digestibility, but also on its biological value, i.e. its capacity of permitting the protein synthesis from food amino-acids.

The amino acids which are necessary for proteinogenesis must be available at the same time in sufficient quantities and precise proportions on the spot of the synthesis, i.e. the ribosomes and polysomes.

Amino acids that the human body cannot synthetise in sufficient quantities are called essential amino acids. There are eight of these : lysin, methionine, leucine, isoleucine, threonine, valine, tryptophane and phenyl-alanine. For the young child histidine is also required. It is known that tyrosine and cystine may partly (40 to 60 %) replace phenylalanine and methionine. These two amino acids are called semi-essential. Table 2 gives an indication of the amino acid requirements in man.

TABLE 2

Essential amino acid requirements (mg/kg bodyweight)

Amino acids	4-6 months baby		10-12 years child		20 years to adult		mg/g of ideal protein (FAO)
	USA	FAO	USA	FAO	USA	FAO	
Histidine	33	28	–	–	–	–	(22)
Isoleucine	83	70	28	30	12	10	40
Leucine	135	161	42	45	16	14	70
Lysine	99	103	44	60	12	12	55
Methionine + Cystine	49	58	22	27	10	13	35
Phenylalanine + Tyrosine	141	125	22	27	16	14	60
Threonine	68	87	28	35	8	7	40
Tryptophane	21	17	4	4	3	3.5	10
Valine	92	93	25	33	14	10	50

Cystine : semi-essential amino acid, can replace 40-60 % of metionine
Tyrosine : semi-essential amino acid, can replace 40 % of phenylalanine.

It is known that proteins of animal origin are better balanced than vetegable proteins to satisfy human requirements. The biological value of animal proteins is superior to that of vegetable proteins because they bring more lysine and methionine which often limit the biological value of vegetable proteins.

This is why lysine is the limiting factor of cereal proteins. The proteins of Leguminosae are particularly deficient in methionine.

As a result cereals are efficaciously supplemented by small quantities of proteins rich in lysine of animal and leguminous origin. Yeasts and some algae constitute an important input of lysine and can advantageously replace animal proteins in the supplementation of cereals.

The following table from J. Mosse and G. Fauconneau (1973) gives an idea of the value of some food proteins, as well as their limiting amino acids.

TABLE 3

Composition of some proteins and their digestibility.

Class of proteic foods	Protein contents % of dry matter	Digestibility coefficient	Limiting amino acid
1. Milk	25 – 35	95	Cystine
Egg	45	95	none
Meat	80 – 95	92	Methionine
Fish	80 – 95	92	Methionine
2. Peas	30	87	Methionine
Beans	30	86	Methionine
Yeasts	60 – 70	86	Methionine
Soya cake	50 – 55	85	Methionine
Peanut cake	50 – 55	84	Lysine, methionine, threonine
Wheat	10 – 14	82	Lysine
Maize	8 – 12	82	Lysine
3. Sunflower cake	45 – 50	78	Lysine
Beans	27 – 30	78	Methionine
Rape seed cake	35 – 37	60	none

The balance of aminoacids in eggs is considered as ideal. Milk also has a good balance. A joint FAO/WHO expert committee (1973) has proposed a standard combination of amino acids which corresponds to the ideal protein, with optimal efficacy and which can best satisfy human requirements at various ages.

2.   The absence of toxic compounds : As said earlier, public opinion is worried by the possible presence of toxic compounds as a result of the use of modern techniques for both plant and animal production (herbicides, insecticides, fungicides for the former and additives, hormones, antibiotics for the latter). There is also the long list of additives incorporated by the industry to preserve food, to correct the loss of taste, smell, consistence and colour.

In fact in each food there are undesired "foreign substances" in addition to the nutrients. But a distinction must be made between the toxic compounds which are present naturally and the additives which have been incorporated with a specific purpose in mind and whose concentration is known, authorized and controllable.

Amongst the industrial additives for human food, let us mention the following :
- preservative agents which are included to avoid a proliferation of bacteria and toxic molds, for example NaCl, sulfites, nitrous salts, CO_2, propionic acid, sorbic acid, sodium benzoate, diphenyls;
- colouring agents, which used to be of natural origin in the last century (saffron, beet red, caramel) but which are increasingly replaced by synthetic products (azoic derivates and tartrazines). Some are not harmless. The lists and authorized doses are regulated, but vary from one country to the other. Prudence is necessary particularly with the azoic colouring agents;
- antioxidants which may either be natural (Vitamins E, Vitamin C), or synthetic (propyl and octyl gallate, dodecyl, BHT, BHA) which are added to prevent fat rancidity;
- acidifiers : citric, lactic, tartric acid;
- synthetic sweeteners : saccharine and cyclamates which are prohibited in many countries - other dipeptides are being tried (Aspartame);
- aromas : about 1,000 are used and mostly of synthetic origin (75 %);
- emulsifiers, stabilizers, thickening and setting agents of natural or synthetic origin.

In general the consumer is protected by legislation, but is it always respected ? Is the control always exerted ? These questions are often raised and this is at the origin of the flourishing industry of the "biological, natural" products.

In fact, as Staron emphasized (1980), there are toxic substances in practically all the traditional raw foods for man, whether they are of vegetable origin (fruit, vegetables, seeds) or of animal origin (fish, shellfish, milk, eggs, meat). These are generally in low doses, but can sometimes be found at higher doses. These toxins may cause several troubles of digestive or metabolic origin (Ferrando, 1983).

To name but a few (cfr. Gontzea et al, 1968 and Ferrando, 1979) :

1. protease inhibitors in several Leguminosae and other families. Fortunately, most of these are thermolabile (antitryptic factors);

2. hemagglutinines or lectines from several Leguminosae (soya, beans, castor bean, peas, lentils) which are also destroyed by cooking;

3. lathyrogens from Lathyrus type beans (Sativus hirsitus, etc...)

4. toxic glycosides amongst which the goitrogens from Cruciferae (isothio-cyanate and 5-vinyl-oxazolidone) and the goitrogens from soya and peanuts which affect the metabolism of iodin and the thyroid gland;

5. cyanogenous substances met in a large number of seeds (various beans, peas, manioc, bitter almonds) which under the action of certain enzymes may liberate toxic HCN;

6. saponins (triterpenoid alcohol glycosides) from soya, beans, peas and peanuts possessing haemolytic properties and inhibiting enzymes and which are also largely destroyed by cooking;

7. isoflavone glycosides from Leguminosae (soya, peas, lupin) and oleaginous plants (rape) which have oestrogenic properties causing an hypertrophy of the organs and retarded growth;

8. neurotoxic substances from various fish species which are accumulated in their ovaries, skin and intestine and which provoke vomiting, paralysis and even death;

9. substances of beans and peas which cause favism (haemolytic disease which lowers the level of glutathion in the erythrocytes);

10. substances which link cations (Ca, Mg, Zn, Cu and Fe) : phytic acid, oxalic acid of spinach, sorel, egg conalbumine;

11. the antivitamins : avidin found in the white of egg is an antivitamin H, citral found in the essential oil of oranges is an antivitamin A, niacinogen found in maize is an antivitamin PP, peas and beans have an antivitamin D, many vegetables and oleaginous plants have enzymes that hydrolyse vitamins (thiaminase, lipoxygenase);

12. allergenic substances which cause digestive or skin lesions and are found in cereal seeds, fruit, etc.;

13. certain ions such as nitrates reduced into nitrites by micro-organisms and nitrites found in lettuce, spinach, or added as preservatives may in the stomach form nitrosamines with the secondary amines (fish and meat). Their carcinogenic action is now certain.

If this list has been presented so extensively, it is to show that even biological and natural products contain toxic substances and an adequate technology is necessary to eliminate these.

There are also toxic substances which are brought by contaminants : this is food pollution. This is the most dangerous group and originates in the treatment channel of plants and animals or appears during the transformation or distribution of the food via the pollution of the environment. Amongst the contaminants, we have several groups :
- the phytotherapeuticals used for vegetable and flower culture where the use of pesticides has always been controversial, but well regulated;
- residues in food of animal origin as a result of phytotherapeutic, veterinary treatment (antibiotics, antimicrobial substances) or the use of additives in industrial feedingstuffs (cfr. Ferrando, 1983). A very strict regulation exists in this respect and ensures the consumer's safety;
- mycotoxins produced by many molds which contaminate foods and the list of which becomes longer each day (Ferrando, 1983);
- polycyclic aromatic carbohydrates which may be absorbed by plants.

But besides these groups, intoxication may be possible by many other routes (Staron, 1980) :
- toxic substances brought accidentally by micro-organisms and false foods;
- toxic products after activation by light;
- parasitic contaminants;
- microbial contaminants (pathogenic germs);
- physical and chemical pollutions (radiation, industrial muds, polychlorates, heavy metals);
- toxic substances which are brought by or appear during technological or cooking treatments.

Conclusion

If we have discussed both the nutritional quality of foods and the problems association with natural potentially toxic substances, this was mainly to help the consumer understand that the survival of our modern society must be based on food production which, itself, is based on the justified and controlled usage of all the factors which permit the supply of quality food at a low price while ensuring the consumer's safety.

Indeed it is too often forgotten that increasingly precise analytical methods detecting residues down to the ppb (part per billion) enable controls to be performed. In addition, is it not reassuring that recent analytical enquiries on the quality of food in several countries such as the United States, the United Kingdom, Japan and France indicated that, when there is an accident, it is often due to a bad usage of the products by insufficiently educated technicians ? Let us add also that "natural" is not always synonymous with "quality" and "health". When it is known that over 70 % of products go through industrial transformations before reaching the consumers' table and that a precise control was possible then, it can be appreciated that there is a system of controls all along the food chain to ensure the protection of human health. Is it not true to say that population in industrialised countries has never been as healthy and as well fed as it is now ? But there is always the risk that man, who is responsible "in fine" of his menu and his food behaviour, does not only follow internal physiological stimuli but is also influenced by the external sensorial stimuli of an olfactive, visual, gustative, tactile and mechanical nature which make him take meals that far exceed what is nutritionally required. One can easily guess the consequences of this : stoutness, cardiovacular diseases and cancers. Excess is bad in any thing. Sociocultural habits and the symbolic value that one gives to food constitute a sort of screen masking their nutritional properties. We must alert public opinion to this.

In our society of abundance, it is high time that the mass be better informed, better educated on a nutritional point of view. Is it not unfortunate to see that neither written press nor television teach consumers that their feeding habits weigh much more in the occurrence of diseases than the undesirable substances which may be present in the food ?

A serious, independent and honest information on the nutritional balance would be necessary at school and at home. Public opinion must be sensitized to "eating right". It must be informed punctually and objectively on all the problems of the production and distribution channels for food.

It is high time that written press and television abandon the path of nihilism. The consumer is always informed of what is "bad" and it is always "somebody else's fault".

Instead of sensitizing the opinion to the toxicological or microbiological risks, consumers' associations should prepare positive information on nutritional balance and the nutritional value of food, sensitize personal responsibility in the "judicious" choice of a "careful diet" which belongs to a healthy hygiene of living. Nutritional education is everybody's concern. The late Professor Tremoliere had entitled one of his last books : "Dietetic is an art of living."

Rather than being a sad discipline that forbids or dissuades, it must become an art of cultivating "health". Let us be conscious of what we eat. Did Brillat-Savarin not say : "Tell me what you eat and I shall tell you who you are" ? Let us fully assume responsibility for our health and that of our family and remember the definition that Hippocrat gave in his time : "Dietetic enables healthy people to remain healthy and helps those who are unhealthy to recover health."

References

A.P.P.R.I.A., 1981. Colloque : Diététique adulte et consommateur moderne.
   Paris, 16 octobre.
A.P.P.R.I.A., 1981. Symposium Européen : La gestion de la qualité des
   produits alimentaires. Paris, 26-27 novembre.
Brisson, J., 1982. Lipides et Nutrition Humaine. Les Presses de l'Université
   Laval-Masson, Paris.
Commission des Communautés Européennes, 1980. Les additifs alimentaires et le
   consommateur. Ed. Office des Publications Officielles des Communautés
   Européennes, Luxembourg.
den Hartog, Hautvast J.G., den Hartog, 1978. Nieuwe Voedingsleer. Uitgeverij
   Het Spectrum, Antwerpen, 7de ed.
Devis, R., 1981. Les prostaglandines : molécules physiologiques ou "étrangè-
   res" ? J. Pharm. Belge, 36, 43-72.
Dupin, H. et Rouaud, C., 1983. L'alimentation des Français. La Recherche,
   146, 964-972.
Dupin, H. et al, 1984. Apports nutritionnels conseillés pour la population
   française. T.E.C. of D.O.C. Lavoisier, Paris.
Ferrando, R., 1973. La qualité des productions animales. In : l'Elevage,
   N° hors série, p. 25.
Ferrando, R., 1979. Les aliments traditionnels et non traditionnels. Coll.
   FAO. Alimentation et Nutrition N° 2, Rome.
Ferrando, R., 1983. Qual. Plant. Plant Foods Hum. Nutr. 32, 455-467.
Ferrando, R., 1983, Quelques réflexions sur la zootechnie et la santé de
   l'homme. Les Elevages Belges, Decembre, pp. 4-5.
Gentils, R. et Jolivet, P., 1979. L'équilibre alimentaire. Flammarion,
   Paris.
Gontzea, I., Ferrando, R., Sutzesco, P., 1968. Substances antinutritives
   naturelles des aliments. Vigot Fr. Edit. Paris.
Langley, Danysz, P., 1983. Cancer : les risques de l'alimentation. La
   Recherche, 150 : 1564-1574.
Massart, D.L., Deelstra, A. en Hoogewijs, 1980. Vreemde stoffen in onze
   voeding. Uitg. De Nederlandse Boekhandel, Antwerpen.
Mosse, J. et Fauconneau, C., 1973. La Recherche : 635-643.
OMS/FAO, 1973. Besoins énergétiques et besoins en proteines. OMS Rome, Rap-
   port technique n° 522, 123 pp.
Rapports du Comité Scientifique de l'Alimentation Animale de la Commission des
   Communautés Européennes :
   (1979) Première série, n° catalogue CB - 28-79-277.
   (1980) Deuxième série, n° EUR 6918
   (1981) Troisième série, n° EUR 7383
   (1984) Quatrième série, n° EUR 8769
   Ed. Office des Publications Officielles des Communautés Européennes, Luxem-
   bourg.
Staron, T., 1982. L'Alimentation Humaine. A.P.P.R.I.A. Librairie Lavoisier,
   Paris (2 vol).
Tremolière, J., 1976. Diététique et art de vivre. Ed. Seghers, Paris.
Tremolière, J., Serville, Y., Jacquot, R. et Dupin, H., 1980.
   Manuel d'alimentation humaine. Tome I : Les bases de l'alimentation.
   Tome II : Les aliments. Ed. E.S.F., 17 rue Viete, Paris.
Vanbelle, M. et Meurens, M., 1984. Besoins nutritionnels humains et qualité
   diététique des aliments. 4ème Cycle : Agro-Alimentation et Biotechnologie,
   A.I.A.L.U. et A.I.G.X., U.C.L. Louvain-La-Neuve, Belgique.
Vandepitte, W., 1983. Ons dagelijks brood : de kwaliteit van onze voeding.
   Belgische Boerenbond. Dienst Public Relations, Leuven.

RESUME

LA QUALITE DES ALIMENTS ET L'ALIMENTATION HUMAINE

Marcel VANBELLE,
Faculté des Sciences Agronomiques,
Laboratoire de Biochimie de la Nutrition,
U.C.L.-Louvain-La-Neuve, Belgique,
Membre du Comité Scientifique de l'Alimentation Animale
des Communautés Européennes.

La notion de qualité des aliments est une des préoccupations essentielles de nombreuses associations pour la défense du consommateur. De tout temps, l'homme a dû se nourrir pour se maintenir en vie, pour assurer ses fonctions physiologiques essentielles; mais avec l'évolution des productions animales et végétales et surtout avec l'amélioration du niveau de vie, la demande pour une diversité dans le choix des aliments s'est largement accrue. La productivité de l'agriculture s'est largement améliorée depuis la seconde guerre mondiale, grâce aux progrès sensationnels réalisés dans les différents secteurs qui la touchent. Tout cela a entraîné une évolution dans la signification même du mot "qualité".

Les méthodes analytiques de plus en plus perfectionnées assurent la protection du consommateur. De plus, quand on sait qu'avant d'arriver dans l'assiette du consommateur, 75 % des produits passent par les industries de transformation, où un contrôle précis peut s'effectuer, on se rend compte que, tout au long de la chaîne alimentaire, un système de contrôle veille sur la santé de l'homme. Mais, le danger existe que l'homme, responsable "in fine" de la composition de son menu et de son comportement alimentaire, ne suivant pas uniquement les stimulations internes d'ordre physiologique, se laisse trop souvent guider par des stimulations sensorielles externes qui font que l'amplitude des repas pris repousse les limites de la satiété au-delà de la satisfaction des besoins nutritionnels. On devine les suites fâcheuses qui le menacent : obésité, maladies cardio-vasculaires, cancers.

Dans notre société d'abondance, il est grand temps que la masse soit mieux informée, mieux éduquée sur sa nutrition. Il faut sensibiliser l'opinion publique à "manger juste". Il faut l'informer de manière ponctuelle, mais objective, sur tous les problèmes qui se posent dans toute la chaîne de production et de distribution des aliments. Assumons la responsabilité de notre santé et de celle des membres de notre famille et retenons la définition qu'Hippocrate donnait en son temps : "La diététique permet aux bien-portants de conserver la santé et aide ceux qui l'ont perdue à la retrouver."

ZUSAMMENFASSUNG

DIE QUALITÄT VON LEBENSMITTELN TIERISCHEN URSPRUNGS

Marcel VANBELLE,

Faculté des Sciences Agronomiques,

Laboratoire de Biochimie de la Nutrition,

U.C.L.-Louvain-La-Neuve, Belgien

Die Qualität unserer Nahrungsmittel ist eines der Hauptinteressen zahlreicher Organisationen zum Schutze der Verbraucher. Seit jeher mußte der Mensch Nahrungsmittel zu sich nehmen, um sich am Leben zu erhalten und um seine wesentlichen physiologischen Funktionen zu gewährleisten, doch mit der Entwicklung der Tier- und Pflanzenproduktion und vor allem mit der Verbesserung der Lebensbedingungen nahm auch der Wunsch nach Abwechslung und einer größeren Auswahl an Nahrungsmitteln stark zu. Seit dem zweiten Weltkrieg hat sich die Produktivität der Landwirtschaft dank sensationeller Fortschritte auf den unterschiedlichsten Gebieten stark verbessert. Dies alles hat auch dazu geführt, daß sich selbst die Bedeutung des Wortes "Qualität" weiterentwickelt hat.

Die analytischen Methoden, die immer weiter vervollkommnet werden, sind für den Schutz des Verbrauchers da. Wenn man weiß, daß 75 % der Produkte – bevor sie auf dem Teller des Verbrauchers landen – von Industrien verarbeitet werden, in denen eine genaue Kontrolle stattfindet, so muß man sich darüber Rechenschaft ablegen, daß das Kontrollsystem in der ganzen Nahrungskette die Gesundheit des Menschen gewährleistet. Es besteht aber die Gefahr, daß der Mensch, der für die Zusammensetzung seines Menüs und für sein Nahrungsverhalten selbst verantwortlich ist, nicht nur den inneren physiologischen Stimuli folgt, sondern sich zu oft von äußeren sensorischen Reizen leiten läßt, was zur Folge hat, daß die aufgenommene Nahrungsmenge die Grenze der Sättigung und des eigentlichen Nahrungsbedürfnisses übersteigt. Die Folgen dieses Verhaltens sind leicht auszumachen : Fettleibigkeit, kardiovaskuläre Krankheiten, Krebs.

In unserer Überflußgesellschaft ist es höchste Zeit, daß die großen Massen besser über die Ernährung informiert werden. Die Öffentliche Meinung ist im Hinblick auf ein richtiges Eßverhalten zu sensibilisieren. Das Publikum ist in punktueller, objektiver Weise über alle Probleme zu informieren, die sich in der gesamten Produktionskette und der Verteilung der Nahrungsmittel stellen. Nehmen wir die Verantwortung für unsere Gesundheit und die der Familienmitglieder auf uns und beherzigen wir die Definition von Hippokrates : "Die rechte Ernährung erlaubt es den Gesunden, ihre Gesundheit zu erhalten, und hilft den Kranken, die Gesundheit wiederzuerlangen."

RESEARCH AND DEVELOPMENT OF A NEW PRODUCT

R. Marsboom, D.V.M.,

Janssen Pharmaceutica, 2340 Beerse, Belgium

In spite of the enormous progress made over the last decades, there is still no entirely satisfactory solution for many serious problems in human, veterinary or plant medicine. Mankind is in need of better and safer therapeutic, prophylactic and diagnostic drugs and is looking toward the scientific community as a whole and toward the pharmaceutical industry in particular to provide these innovative products.

New products for animal and plant health care are developed by the pharmaceutical industry with - but mostly without - the help of institutes. On the one side, such new products are applied in companion animals (pets, equines) who play an increasingly important role in the psychological welfare of the young and the elderly in the Western world. On the other hand, it is well recognized that, since food demands and food needs for the growing world population are steadily expanding, there is an increasing market need for products improving livestock productivity, meaning more and higher quality food of animal origin (meat, milk, eggs). The basic determinant of livestock systems with intensive stock husbandry is the use of confinement feeding as a major production technique. This feature shapes a demand for new and better preventive and therapeutic products like vaccines, antimicrobials, antifungals, endo- and ectoparasiticides, coccidiostats, growth promoters, etc. in various species, but more specially in cattle, pigs and poultry.

The research and development function at the industry level envisages the future as one of great opportunity, but with increasingly greater responsibilities toward mankind. Improved medicines in animal health are the result of pinpointing the shortcomings of the existing products and of an accurate definition of the customer's needs for an optimal and economically profitable livestock production.

New products will not result anymore from "haphazard" findings. Research and development activities are necessarily based upon a cohesive and responsible research and development programme which means 99 % perspiration after 1 % inspiration. Furthermore, successful drug research requires a vital inner structure (teamwork, enthusiasm and motivation) and an unstiffing outer structure. The point I want to make in this connection is that if society really wants its essential health needs and food demands to be met fully and quickly, it should create a proper motivating climate in which creative drug research is likely to prosper.

1.   The daily perspiration

It is definitely not possible to describe from conception to production or from brain to bottle the various steps in the development of a new product. The crucial bottlenecks are :

a) <u>a clear definition of the needs</u> : there is, of course, a steady evolution in the animal production industry and thus an ongoing change in the need for various types of products. Searching for another broad-spectrum anthelmintic in a given class of chemicals is unrealistic, since this market is already oversupplied and suffers from price cuts of existing anthelmintics. Research for a new anthelmintic should, therefore, be directed towards low-dose active compounds or products with better activity against e.g. ostertagiasis or broad-spectrum products with endo- and ectoparasiticidal effects.

What are the major needs in livestock production ? The following issues are fundamental :
- improvement of husbandry and management;
- improvement of productivity;
- enhancement of immunity.

Specific diseases that continue to exist are due largely to poor management and husbandry, i.e. mastitis, MMA, the calf pneumonia-diarrhoea complex, baby-pig diarrhoea, mycotic infections particularly in poultry, some endo- and ectoparasitic diseases, foot rot, some forms of infertility. The pharmaceutical industry can provide drugs to substitute partly for poor management, although this is certainly not the best or final answer. However, since good husbandry and management cannot be imposed, it is just plain logic that the pharmaceutical industry has active research and development programmes in the search for new vaccines, antimicrobials, antifungals for skin forms, intestinal forms, aspergillosis and deep mycoses, coccidiostats, endo- and ectoparasiticides, growth promoters.

Improving productivity and enhancing resistance or immunity are not easy challenges. The pharmaceutical industry can play a vital role in such development by learning more about biology and the mechanism of action of growth promoters and immunomodulators. If we know the mode of action of actually used products, new and more potent products will be easy to formulate. Non-hormonal and non-antibiotic compounds in growth promotion are already in development for various species.

b) <u>An open eye and ear</u> will not only allow for a clear definition of the needs but may also lead to new applications of existing compounds. The observation in the 70's that the host defense mechanism against infectious disease increased after the use of the anthelmintic levamisole resulted in its use as an immunomodulator in some chronic infectious diseases, cancer and systemic diseases.

c) <u>Chemical hat and coat stand</u> : once the needs and goals have been defined, synthesis of new molecules necessitates a chemical hat and coat stand on which one can hang the new planned molecules. This can be done either the empirical way or preferably the retrosynthetic way, whereby new molecules are first designed on paper and later synthetized.

d) <u>Experience in pharmacological chemistry</u> : once the definition of the course(s) of synthesis has been made, it requires a lot of skill and experience in pharmacological chemistry to transfer the new molecules into pharmacologically applicable products. Substitution on molecules by fine technology will allow to :
- avoid penetration in the central nervous system (CNS),
- assure fast metabolism and excretion through faeces or urine,
- guarantee the absence of residual levels in food chain animals so that withdrawal periods can be established for safe human consumption, but remain compatible with economic considerations.

e) <u>Pharmacological screening</u> : several thousands of compounds have to be synthetized before one is retained for further development. This requires pharmacological screening using either standard tests or new screening models. The ideal swivel bridge in pharmacological screening does not exist and new methodology has often to be developed in order to find new drugs. For a long time pharmacologists have been convinced that a periodical review of their routine testing procedures is necessary. This must include a critical evaluation of the performance of the various tests, a detailed analysis of results and incidents in which the animal tests predicted a pharmacologic response that failed to materialize in the target species. The lack of success in screening generally relates to the lack of representative research models.

From chemical synthesis to pharmacological screening, and toxicity and kinetic studies, the clinical safety and efficacy can be determined, resulting after registration into a new drug application. Only one out of 8,000 newly synthetized products will become a new medicine. Such development is clearly a time-consuming operation involving an investment of about 50 million US dollars or more per single new drug.

f) <u>Safety control</u> : the safety of new molecules is sometimes more important than efficacy. The main purposes of toxicity testing are :
- to show the circumstances in which a compound has an unwanted or harmful

effect in the laboratory, e.g. dose, duration of treatment, route of administration, etc.;
- to try to understand the mechanism of that action;
- to predict the likelihood, nature and severity of any adverse reaction in
patients (man, animal) in relation to a proposed medical treatment;
- to decide whether the harmful effect of the disease to be treated justifies
the risk of the proposed therapy.

Many toxicologists are resented for their obedience to rigid regulations that interfere with the scientific flexibility necessary for drug development. However, toxicity testing should not be considered nihilistically
but as a real standard of value in the development of better and safer medicines. The benefits should be weighed versus the adverse effects.

Toxicological testing of pharmaceuticals has become a huge enterprise
in which industrial, governmental, academic and private laboratories are
involved and for which enormous sums are paid every year. Costs today for a
full battery of toxicity tests are estimated to be around 4 million US dollars. It is important that the concepts of testing and the techniques applied
are regularly and critically reviewed mainly with regard to scientific base,
cost effectiveness and ethical justification.

One should also remember that any modern therapeutic interferes with a
biological process and thus by definition modern medicines are all potentially
toxic. Perfect safety is thus a chimera and overregulation should not
strangle human activity in a search for the impossible.

g) <u>Pharmacokinetic testing</u> of products used for the improvement of
livestock productivity is as important as efficacy and safety. This is
specially the case for products administered to food chain animals whereby an
accurate understanding of metabolism, excretion and residual levels is vital
for the determination of the acceptable daily intake by man at indicated withdrawal periods. This will guarantee safe food to the consumer.

h) <u>Clinical experimentation</u> : a traditional complication in drug development is the phase of clinical experimentation when confirmation of efficacy
is sought outside the laboratory under field conditions. Almost by definition
clinicians are without discipline and any clinician should swear fidelity to
the experimental protocols if one wants to obtain reliable results. Indeed
the finest computer and the best statistician are helpless when incorrect or
incomplete data are provided.

2.   <u>The importance of a vital inner structure</u>

The pharmaceutical industry has the moral obligation and responsibility to provide mankind with suitable and safe products. The lesson we have

learned over the past decades is that biological research cannot be segregated into discreet components. Such a philosophy strengthens the basic concept that there is only one research as well as there is only one medicine with human, veterinary and even agricultural components. These components are influenced by the sociologic and economic climate of the various geographical areas of the globe.

Only innovative products are the driving force of industry. If a company survives today only because of its inherited momentum of the past, this provides an uninspiring environment in which to work. Novel products must be of a high scientific performance, contribute to progress and have a chance of success in a foreseeable future. A vital inner structure is essential for achieving the required good results. It is built upon a quadruple foundation :

a) A multidisciplinary endeavour : research and development at industry level are like an extension of the university campus. One encounters chemists, biochemists, biologists, pharmacologists, toxicologists, pathologists, clinicians, statisticians, engineers, economists and many other experimental scientists. Contrary to university where research workers sometimes limit themselves to their own areas of interest, research people in industry live and work together in a group as a team.

The pharmaceutical industry is not a monolithic block, i.e. all companies are not the same and each tries to be personal. But the common basic research organization within industry is invariably the same : from literature over synthesis and analysis to pharmacology and toxicology, and then development and marketing in one or more of the three health care sectors. It only seems evident that any such highly multidisciplinary staff should work with the motto : "Uniformity kills and variety gives life".

Conducting such a team is not an easy task. It requires perspective, a deep understanding of the problems to be solved together, enthusiasm and, above all, motivation before the ideal wonder drugs of the future can become reality.

b) Utilization of specific knowledge and skill : the second pillar for a vital inner structure within industry is the utilization of the staff's specific knowledge and skill; in other words, projects are built around men and not vice versa. It is not a coincidence that some industries do better than others in a particular area. The only reason for this is the commitment of experienced and qualified scientists for that area. Some industries are specialized in vaccines, others in antimicrobials, in tranquillizers, or in endo- and ectoparasiticides and so on.

c) Fundamental research : the most vital pillar is room for fundamental research. Needless to say that the development of "me too" products will soon become superfluous, even unwanted, since the world will have no further need for only variations on already existing products.

We must ensure that some of our resources (people, time and money) are made available for fundamental projects. If we want to open the way to untrodden paths, we must stop believing that we are too poor to do fundamental research. Rather than hiring someone to develop a vaccine against a given disease, it is better to hire someone to study the organism, to learn its innermost secrets. Once we know the secrets, the vaccine will simply be a natural by-product of the research programme.

The challenge of innovative research programmes can only be met with the full cooperation of academias and the governments under which we operate. Because the diseases or problems we are trying to solve become increasingly sophisticated, we must form a compatible research structure that attacks these problems, shares knowledge and supports bright scientists and their ideas while being responsible and responsive to the needs of society.

d) <u>A limited hierarchy</u> : the fourth pillar for a vital inner structure is "freedom". There must be room for creative thinking, for an open, intensive and informal communication between research workers within a team and between teams.

Therefore, it is extremely important that horizontal structures exist within the pharmaceutical industry with ways and means of direct communication. In case of over-organization and over-regulation occurring incipiently but surely in each growing organization, bureaucracy will soon suffocate creative thinking and innovative realizations.

3.   <u>The danger of a morbid outer structure</u>

What has been said with regard to the inner structure is also applicable to the outer structure or to the environment in which the pharmaceutical industry has to operate. Many factors from outside indeed influence dramatically the research and development activities of the industry and explain why it takes so long to develop new drugs.

Let me summarize in a few words the classical history of most modern drugs. Shortly after the announcement of its discovery the new drug often tends to be ignored or met with sterile scepticism : somebody is trying to sell something. The world has been fooled so often in the past, so let us be extremely careful this time and wait for independent confirmation, for more data, more facts, more double and triple blind experiments, absolute proof of safety etc.

In a second phase, when the obvious facts can no longer be denied, the general feeling changes : there must be something to it. At this stage, however, the usual rumours start spreading about all kinds of mysterious dangers. Smoke is being produced and people start looking for a fire. "This new drug, my friend, is much too toxic to treat your cat with."

It is evident that since a new drug is made available to the public, the new drug in fact does not belong anymore to industry. Society, via the authorities, rightfully imposes high standards on new drugs, but the requests for e.g. guarantee of safety are not always based upon a rational concern for evaluating risks, but merely upon an unrealistic fear for the unknown. Furthermore, the political commitment to health care has isolated the pharmaceutical industry in many countries by increasing requirements on safety, toxicity, efficacy, stability, coupled with more guidelines regarding presentation of data and, therefore, resulting in more time-consuming and more costly processes to obtain new drug applications. In a "unified Europe", old-fashioned import and export barriers are disappearing but become replaced by more bureaucracy in chauvinistic nationalism. We are indeed living in a strange world with tremendous scientific and technological potential on the one hand, and alarming indications of an even more impressive, rapidly growing but suffocating dictatorial bureaucracy on the other hand.

My dictionary defines a bureaucrate as an official who works hard, but only obeys the rules of his department without exercising much judgement. It is to this extremist bureaucracy that I am referring, as we are all aware of, accept and even want the existence of governmental regulations.

Just to give an idea of the demanding requirements industry has to face : in a typical year the FDA conducts 32,000 inspections, performs 98,000 wharf examinations, 50,000 sample examinations, issues 300 regulatory letters, initiates 500 product seizures, 40 injunctions, 15 prosecutions, monitors 1,000 product calls, receives 33,000 FOl requests, responds to 55,000 consumer inquiries, averages 40 congressional hearings, receives more than 2,500 press inquiries (FDA Consumer, June 1981).

Comparable figures exist for other countries and I do not question the need for such procedures. The question is : Were they all constructive, do they all serve useful purposes ?

A deplorable conflict is developing between researchers and large segments of the population, merely created by the rapid intrusion of new and revolutionary technologies in everybody's life. Criticism, scepticism, suspicion and even hatred as they are conveyed by certain mass media to part of the population, deserve attention but should not determine the lines of conduct since criticism usually does not lead to renovation or improvement. Such challenges can be met with full cooperation between industry, academias and the governments.

The social goals and scientific innovation of new therapeutics as well as their economic benefits to mankind should be communicated openly to the public who has a real desire to understand better and also the right to be clearly and fully informed.

Finally, the Patent Act of 1836 guarantees any inventor the exclusive rights to exploit his invention during 17 years, constituting a protection against illegal competition and piracy. Since the period for the development of a new drug steadily increases and in some cases even exceeds the patent protection period, the enforcement and prolongation of patent protection for innovative products would encourage industry to invest more in research and development of safe and useful new products.

Conclusion

The pharmaceutical industry envisages the future as one of great opportunity with more difficult challenges and with a greater responsibility toward mankind. If society really wants its essential health needs to be met quickly and fully it should try to create a proper motivating climate in which creative drug research and development is likely to prosper. These challenges can only be met with the full cooperation of academias and the governments under which industry must operate.

RESUME

RECHERCHE ET DEVELOPPEMENT D'UN NOUVEAU PRODUIT

R. Marsboom,
Docteur en Médecine Vétérinaire

Malgré les nombreux progrès réalisés ces dernières années, des médicaments thérapeutiques et prophylactiques nouveaux et meilleurs continueront à être développés afin d'améliorer la santé animale et de procurer à l'humanité de plus grandes quantités d'aliments d'origine animale de qualité supérieure. Notre société s'attend à ce que l'industrie pharmaceutique lui fournisse ces produits meilleurs et plus sûrs.

Le futur représente, pour la fonction de recherche et développement de l'industrie, une source de nouvelles ouvertures, mais assortie de plus grandes responsabilités vis à vis de l'humanité. L'amélioration des produits de santé animale est le résultat d'une meilleure définition des besoins des consommateurs pour une production animale optimale et économiquement rentable.

Le premier stade consiste en un repérage des carences des produits pharmaceutiques existants. Comme les nouveaux produits importants ne proviendront pas de découvertes "accidentelles", recherche et développement sont nécessaires et sont effectués sur base d'un programme de recherche cohérent et responsable, exécuté en équipe, dont les stades principaux sont la définition des processus de synthèse, le screening pharmacologique utilisant des tests standards ou de nouvelles méthodes, des études toxicologiques et cliniques chez les espèces visées. La cynétique et la détermination d'un niveau d'ingestion quotidien acceptable en rapport avec les délais d'attente prescrits doivent finalement garantir l'innocuité des denrées pour le consommateur.

Pour l'industrie pharmaceutique, c'est une obligation morale et une responsabilité de mettre à la disposition de l'humanité des produits sûrs et adéquats. Seuls les produits originaux permettent à l'industrie d'évoluer, mais une structure interne est indispensable pour atteindre de bons résultats. Elle repose sur quatre éléments fondamentaux :

1. un staff fortement polyvalent dont la devise est "l'uniformité tue et la variété anime";

2. l'utilisation des connaissances et capacités spécifiques des membres du staff, c'est à dire l'établissement des projets en fonction des hommes et non le contraire;

3. une ouverture vers la recherche pure et la découverte de nouvelles voies en collaboration avec les instances académiques et gouvernementales;

4. une hiérarchie réduite qui permette une communication informelle intensive entre les chercheurs.

Si notre société veut réellement que ses besoins essentiels en santé et en alimentation soient comblés rapidement, elle se doit de créer un climat favorable dans lequel la recherche et le développement peuvent prospérer. Une structure externe souple est donc également indispensable.

Il est juste que, par l'intermédiaire de ses autorités, la société impose des normes élevées pour les nouveaux produits. Mais les exigences, par exemple les garanties d'innocuité, doivent être basées sur un désir rationnel de déterminer les risques et non sur une crainte irraisonnée de l'inconnu. La protection des brevets d'invention devrait être prolongée, car le développement de médicaments s'étale sur de longues périodes qui réduisent la durée des ventes sous brevet. Enfin, la critique, le scepticisme, la méfiance, voire la haine communiqués par certaine presse à une partie de la population, méritent l'attention, mais ne devraient pas déterminer la ligne de conduite. Ces défis ne peuvent être affrontés qu'en étroite collaboration de l'industrie avec les académies et nos gouvernements. Les buts sociaux et les inventions scientifiques de nouveaux produits thérapeutiques, ainsi que leurs avantages économiques pour l'humanité devraient être communiqués ouvertement au public, qui a non seulement le souci de mieux comprendre, mais également le droit d'être informé correctement.

ZUSAMMENFASSUNG

FORSCHUNG UND ENTWICKLUNG AN EINEM NEUEN PRODUKT

R. Marsboom, D.V.M.

Trotz des enormen Fortschrittes in den letzten Jahrzehnten verbessern auch heute noch neuartige und besser therapeutische, prophylaktische und diagnostische Mittel die Gesundheit der Tiere und verschaffen uns höhere Mengen qualitativ hochstenhender tierischer Nahrung. Unsere Gesellschaft erwartet von den Wissenschaftlern und besonders von der pharmazeutischen Industrie, daß sie diese besseren und sicheren Mittel entwickelt.

Forschung und Entwicklung auf industrieller Ebene sehen in der Zukunft große Möglichkeiten, allerdings mit zunehmend wachsender Verantwortung gegenüber de Menschheit. Verbesserte Arzneimittel für die Tiergesundheit widerspiegeln das Bedürfnis des Käufers im Hinblick auf eine optimale und auch in wirtschaftlicher Hinsicht ertragreiche Tierproduktion.

Der erste Schritt besteht darin, daß wir Unzulänglichkeiten der existierende Pharmakotherapeutika genau erfassen. Anschließende Forschungs- und Entwicklungsarbeiten müssen notwendigerweise auf einem zusammenhängenden und verantwortungsbewußten Forschungsprogramm und einem Teamwork gründen, da wichtige Produkte heute nicht mehr durch Zufall aufgefunden werden. Die Hauptstufen dieser Entwicklung sind : Definition des oder der Synthesenverfahren, pharmakologisches Screening mit Standardtests oder neuen Modellen, toxikologische Studien bei der betreffenden Tierart und klinische Erprobung. Die Kinetik des Arzneimittels und die Bestimmung der pro Tag zulässigen Aufnahme durch den Menschen und der Angabe der Absetzfristen müssen schließlich die Sicherheit des Nahrungsmittels für den Konsumenten garantieren.

Die pharmazeutische Industrie hat die moralische Verpflichtung und Verantwortung, die Menschheit mit geeigneten und sicheren Produkten zu versorgen. Nur innovative Produkte stellen die treibende Kraft für die Industrie dar, doch um gute Ergebnisse zu erhalten, muß eine lebensfähige innere Struktur vorhanden sein. Sie steht auf vier Säule :

1. eine ausgeprägt multidisziplinäre Forschergruppe, die unter dem Motto steht : "Gleichförmigkeit tötet, und Vielfalt gibt Leben";

2. die spezifischen Fähigkeiten und Kenntnisse der Forschergruppe müssen ausgenutzt werden, d.h. Projekte werden um den Menschen herum gebaut und nicht umgekehrt;

3. Raum für Grundlagenforschung öffnet den Weg zu noch nicht begangenen Pfaden in Zusammenarbeit mit Akademien und Regierungsstellen,

4. eine nur begrenzte Hierarchie, die intensive und nicht von Förmlichkeiten belastete Kommunikation zwischen den Forschern begünstigt.

Wenn unsere Gesellschaft tatsächlich wünscht, daß der Bedarf nach Nahrungsmitteln und Gesundheit schnell und voll erfüllt wird, sollte sie das geeignete motivierende Klima schaffen, in dem eine kreative Forschung und Entwicklung von Arzneimitteln gedeihen kann. Dazu ist unbedingt auch eine informelle äußere Struktur vonnöten.

Die Gesellschaft auferlegt über ihre Behörden neuen Arzneimitteln zu Recht einen hohen Standard, doch die Forderungen, z.B. im Hinblick auf eine Sicherheitsgarantie müssen auf rationale Abschätzungen des Risikos und nicht auf unrealistischer Angst vor dem Unbekannten beruhen. Kritik, Skepsis, Verdacht und sogar Haß, wie sie von gewissen Massenmedien einem Teil der Bevölkerung vermittelt werden, verdienen zwar unser Aufmerksamkeit, sollten aber die Linien unseres Verhaltens nicht bestimmen. Diesen Herausforderungen kann nur eine volle Zusammenarbeit zwischen Industrie, Akademien und den Regierungsbehörden begegnen, unter deren Aufsicht wir tätig sind. Die sozialen Ziele und die wissenschaftliche Innovation neuer Therapeutika sollten ebenso wie ihr wirtschaftlicher Nutzen dem Publikum offen dargelegt werden, denn es wünscht in der Tat, diese Dingen besser zu verstehen, und hat auch das Recht auf klare Information.

Schließlich würde die Verschärfung und Verlängerung des Patentschutzes für innovative Produkte die Industrie ermutigen, dauernd in Forschung und Entwicklung sicherer und nützlicher Produkte zu investieren.

<u>DISCUSSION</u>

Intervention of Mr. Pierre Romeyer, chef and owner of the restaurant "Maison de Bouche - Pierre Romeyer", Hoeilaart, Belgium.

Gentlemen, I am the link between your production and the public, and there is one thing which should not be disregarded : this is taste, the taste of nature. I must sometimes prepare dishes with ingredients which smell or taste of nothing. You may be the best in your discipline, but you will never manage to prepare a tasty soup with tasteless vegetables. You certainly know of people who grow their own vegetables or keep their own laying hens. Talking about eggs, they are available in large quantities, but they are weak. I used to need eight whole eggs to prepare a "Viennoise", now it takes fifteen eggs to achieve the same result. Eggs have lost their strength; the hens are fed for what they produce and when they become unable to lay eggs, they are slaughtered for the soup. This is just an image.

Society must be protected. But what is society, you or the public ? There are risks : the farmer who is told to give one injection once a week to his animals may be tempted to give one injection per day to increase his profit. This is human.

I have visited slaughterhouses for pigs. The animals are killed, cut in two and steamed. The meat swells in the steam and is filled with water. It is heavier on the scale and profit is better. When I cook this meat - on a grill to prevent it from soaking in its water - it loses one-third of its original weight. This is a choice of society, but who makes this choice ? You, the public ?

I have once been invited to see how apple trees are grafted at the school of Anderlecht which is famous worldwide for trimming and grafting of trees. I have raised the simple question : what is the place of taste in the selection of fruit ? Taste is number six. Low trees are grown to reduce the risk of work accidents and to increase productivity, as more fruit is picked when the worker remains on the ground. The fruit must look nice, be transportable and resistant to natural attacks. Then comes taste. One tonne of "Golden" apples in this room would smell nothing, but six wild apples would perfume the whole room.

Question : It seems that the consumers no longer trust the authorities who approve feed additives, pesticides, substances for plant protection, etc.

What can be done to restore this confidence ?

Y. Domzalski : We cannot say that we do not trust the authorities, but there have been a number of instances when it proved necessary for consumer associations to play their role of watchdogs. One of these was the case of US wines. In July 1983 after several years of negotiations the European Community decided to authorise the importation of wines from the United States. These wines contained a number of substances which were either unused or unauthorised by the European regulations on wine production. The consumer associations have obtained a significant reduction of the number of those substances which could be found in American wines. It should be accepted that, similar to a parliament in a democratic country, the consumer association should play a role of watchdogs and the key to ensuring their confidence in the authorities is to ensure a maximum of transparency in decision making.

Question : Why do consumers so often have bad advocates who hide their incompetence behind their search for spectacular matters ? The hormone matter is an example of this.

Y. Domzalski : I accept that consumer representatives often react aggressively. Let us not forget that consumer associations in Europe are relatively recent (maximum 20 years). They have limited resources and cannot therefore use the services of experts of a similar level as those employed by industry. We work with experts employed by the public authorities. At the E.C. Commission level we sit in expert committees, but only have a consultative voice. Talking about hormones, the Community decided to ban all hormones four years ago. At that time we requested that the expert committees set up to study this matter should include representatives of the consumers but received no reaction. There are no experts from the consumer associations in these committees. Although we do not deny the quality of the experts in these committees, we feel that all interested parties should be represented. We also believe that the Commission tries to delay the approval of some natural hormones (trenbolone and zeranol) in the hope that the consumers will drop the matter and that, with the pressure from the industry being stronger, the artificial hormones will also eventually be approved. This is unacceptable and I want to reiterate that all interested parties should be represented in the expert committees and informed on what is going on.

Question : Regarding the hormone for increased milk production, do you think there are tentative plans to market it and impose its use ?

Mr. Pierre Romeyer.

Prof. Raymond Ferrando,
Member of the Scientific Committee for Animal Nutrition of the European
Communities

Left to right : Dr. Robert Marsboom, Janssen Pharmaceutica; Prof. Marcel
Vanbelle, U.C.L.-Louvain-La-Neuve; Prof. Giovanni Ballarini, Università di
Parma; Dr. Jean Radisson, Ralston Purina(first session moderator);
Dr. Graham Meadows, European Commission; Mr. Yves Domzalski, B.E.U.C.

Left to right : Prof. Marcel Vanbelle; Dr. Jean Radisson; Dr. Graham Meadows;
Mr. Yves Domzalski; Dr. Peter Storie-Pugh,Economic and Social Committee of
the E.C. (discussion leader).

Y. Domzalski : My sources of information are scientific publications issued at the end of last year and comments made to me by E.C. civil servants who discovered that, although it had not yet been marketed officially, this hormone was being used in Denmark and the Netherlands. I guess the companies involved spent a large amount of money in developing this hormone and will naturally try to sell it. Can this occur without consultation ? We may need this hormone in a few years when only $1\frac{1}{2}$ to 2 % of the European population will be working in agriculture, but at the moment I do not think we need it.

Question : What is the effect of guaranteed purchases to the Third-World on surplus production in the E.C. ?

Y. Domzalski : It is paradoxical that imported feedingstuffs are cheaper than locally produced feedingstuffs. This means on the other hand that a number of agricultural surplusses such as milk for example are now no longer used for animal feeding. We can obviously not stop our imports of manioc or soybean from developing countries or the US, but it would be reasonable to look for an equilibrium. This should be discussed at the level of the GATT involving all interested parties.

Question : How can the Commission be sure that the system of milk quotas will have no perverse effect in the sense that milk producers will have to pay enourmous sums of money to buy their quota and will, therefore, pass this high fixed cost on to the consumer ? Why did it not chose to put the milk producers in to the same sort of market regime as those for eggs and poultry, i.e. on the market more or less pure and simple ?

G. Meadows : We did not put milk producers on the market pure and simple because we are starting out with a surplus and this would bring about the greatest possible disorder in the market place. Secondly, milk production is different from the other commodities in that it is carried out on farms and not on what one might call intensive systems. To stop quotas from having a perverse effect we will not make these quotas transferrable freely between producers. Quotas will be tied to the land and in order to buy a quota you will have to buy a farm, and that, we would hope, will slow people down. Talking about animal production in surplus and increased production of animal products through the use of hormones as referred to by Mr. Domzalski, there is a distinction to be made between productivity and production. What we want to do is maximize productivity and at the same time keep production within reasonable limits. After all, it is the tax-payer who has to pay for the disposal of unwanted products.

Question : What are in your opinion the (most) toxic natural substances in human food ?

Prof. R. Ferrando : I have published a 200-pages book for the FAO on the antinutritional substances found in natural food. They are very numerous and a lot of these substances are destroyed by cooking. One finds a number of toxins, biogenic amines and antivitamins in fish, cheese, shellfish, caviar, whortleberries and coffee. There are allergenic substances in a large number of foods. My late colleague, Prof. Fraser, once said at a WHO committee meeting : "It is not because some people are allergic to lobster that this delicious crustacean should be prohibited". Barbecue should be prohibited, because a large number of substances, some of which are terrifying, are produced when grilling steaks. Molds found on untreated fruit produce carcinogenic substances. As you can see, there is an equilibrium and, to conclude, I would wish that man's brain was as tolerant as his body.

Question : Is there not a need to define the term "food quality" as this can be influenced by cooking and other treatments in which it is subjected by the consumer ?

Prof. M. Vanbelle : Quality is a very complex notion, which varies according to micro and macro sociological patterns. Quality is a relative notion which evolves in parallel with the progress made in the production and distribution conditions of goods and also the behaviour of the consumer. Product quality criteria depend on social, economic and psychological factors. From a nutritional point of view, quality includes chemical formulation, energy digestibility, fats, sugars, proteins, vitamins, minerals, and for proteins the important factor is the metabolic availability at the level of ribosomes. In addition it is obvious that food quality changes during the technological processes to which food is subjected.

Question : We eat 20 to 40 % more calories than we need. How do you explain that the majority of people manage to maintain a reasonable balance and consistent weights ?

Prof. Vanbelle : Individual variations are wide. Some people have a very good food conversion ratio and can hardly eat or they put on weight. It appears however that there is a increasing number of people who watch their weight, make more exercise and spend more calories. Nutritional education should teach people to use more calories and eat less. If people ate less, there would be less heart attacks which are usually due to nutrition imbalance and in particular excessive fat intake.

Question : How can you build a church on FAO statistics for developing and industrialised countries ? What are these data on energy intake worth ?

Prof. Vanbelle : FAO data are collected in a number of countries and averages are calculated. These statistics can only be indicative of a trend.

Question : How many years pass as an average from the discovery of an active ingredient to its marketing ?

Dr. Marsboom : This may be very variable depending on the country or type of product. As an average we calculate a number of eight years, but it may be as much as 17 years, long past the patent protection period. It depends on the type of product. When the product is intended for food producing animals, it takes much more time because of the residues studies which must be conducted, but in general we calculate eight years.

SESSION II: SAFETY

Prof. André Rico,
Laboratoire de Toxicologie Biochimique et Métabolique (INRA), Ecole Nationale
Vétérinaire de Toulouse, France.

Prof. Michiel Debackere, Dean of the Rijksuniversiteit Gent, Belgium.

Prof. Paul Delatour, Laboratoire de Biochimie, Ecole Nationale Vétérinaire de Lyon, France.

Dr. Francis J. C. Roe, Independent Consultant in Toxicology, Experimental Pathology and Cancer Research.

Prof. Vivianne Burgat-Sacaze, Laboratoire de Toxicologie Biochimique et Métabolique (INRA), Ecole Nationale Vétérinaire de Toulouse, France.

Dr. David J. Taylor, Department of Veterinary Pathology, University of Glasgow, Veterinary School.

Dr. Cyrille Tancrède, Service de Microbiologie Médicale, Institut Gustave Roussy, Villejuif, France.

Dr. Peter Altreuther, Pharmaceutical Research Centre, Bayer AG, Wuppertal.

VETERINARY DRUGS AND PUBLIC HEALTH

THE GENERAL PROBLEM OF SAFETY

André RICO,

Laboratoire de Toxicologie Biochimique et Métabolique (INRA)

Ecole Nationale Vétérinaire de Toulouse,

F-31076 Toulouse Cedex, France.

Human health is linked directly to the environment and in particular to the nature and quality of his nutrition. Veterinary drugs may represent a concern for public health as a result of the presence of residues in meat from animals which have been treated. As a large number of consumers believe that "natural" means "safe", it is obvious that, in their views, these residues which are usually constituted of synthetic chemical molecules seem to be an important hazard. Consequently, the following points will be addressed in this presentation :

- Does eating even "natural" foods present a risk ?
- What are veterinary drug residues and what do they represent ?
- What is their toxicological significance ?

1.   Does eating even "natural" foods represent a risk ?

There are large numbers of toxic compounds in food (Ames, 1983). Coffee for example contains chlorogenic acid, which produces caffeic acid, a mutagenic phenolic compound, and caffeine which has been shown experimentally to have teratogenic, even carcinogenic, properties. One cup of coffee contains about 200 mg of the former and 100 mg of the latter. Potatoes contain two alcaloïds (solanine and chaconine) which are powerful cholinesterase-inhibitors and possible teratogens (15 mg per 200 g). Black pepper contains small quantities of safrole, a substance which has shown carcinogenic properties in rodents, and practically 10 % of a compound very similar to piperine. Furthermore, extracts of black pepper induce tumours in various organs of the mouse at doses equivalent to the daily intake of a large number of people. Many molds contaminate our food (peanuts, bread, cheese, fruit,...). Some of these molds may synthetize very aggressive mycotoxins such as aflatoxins or sterigmatocystine, which are powerful cancer agents.

It is also well known that cooking (barbecue), smoking or salting may lead to highly mutagenic or carcinogenic compounds (pyrolyzed aminoacids, caramelization, nitrites, nitrates). It has been established that tight links exist between colon cancer and low vegetable diets, rich in meat, between breast cancer of the woman and the fat contents of her food.

In summary it would appear that contrary to the usually admitted beliefs, "nature is not harmless". From a safety point of view, to oppose natural and man-made xenobiotics is a toxicological nonsense. Whatever his origin may be a xeniobiotic remains a foreign agent to the human organism. Lastly, "to take food is to take risks" and the problem is to know if veterinary drugs are likely to increase these risks.

2.   Residues of veterinary drugs

Following administration of a drug to an animal the drug is metabolized. This metabolization aims in fact at improving its elimination and for a large part its detoxification. However, such products can be found in milk and eggs, and also in meat and offals. It must be noted that the drug was administered to the animal at a therapeutic, i.e. little or non toxic dose and it can be observed that the amount of residues is thus very small.

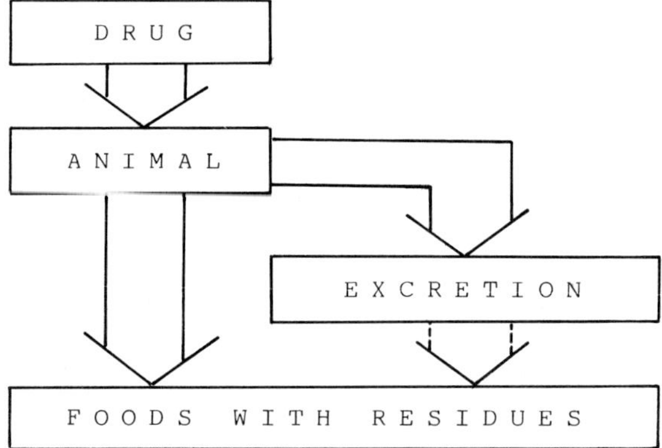

Taking the above into account, it can thus be said that :
- residues can consist of the parent molecule and its metabolites, the latter often being less toxic than the former;
- the levels of residues are low to very low;
- and in fact to quantify these levels, concentration units which are used are the ppm (part per million) = 1 mg/kg, the ppb (part per billion) = 1 µg/kg and the ppt (part per tonne) = 1 ng/kg. When it is said that a substance contains 0.1 ppb of residue X, this means that it contains 0.1 µg/kg or 100 ppt or 100 ng/kg of this residue.

These measures may seem very abstract for the layman, so to give an idea of what they represent :

- one ppm in terms of time corresponds to one minute in one million, i.e. in two years;
- one ppb in terms of volume is one drop in $50m^3$ or in 50,000 litres;
- one ppt in terms of surface corresponds to one 0.50 FF coin lost in the town of Paris. These examples do, I guess, compare quite well the residues as far as their value in meat from treated animals is concerned.

An idea that is often put forward and defended, and which at first sight, appears to be very judicious and satisfactory, is that one should only accept the use of meat from treated animals when all administered drugs have been totally eliminated. One then talks of 0 tolerance.

This notion of 0 tolerance corresponds in fact to the absence of residues. If it was retained some 15 years ago, it has been abandoned as analytical methods developed to make the zero increasingly smaller. This is shown in the schematic curve below, which symbolizes the limits of analytical sensitivity of the ppm, ppb and ppt.

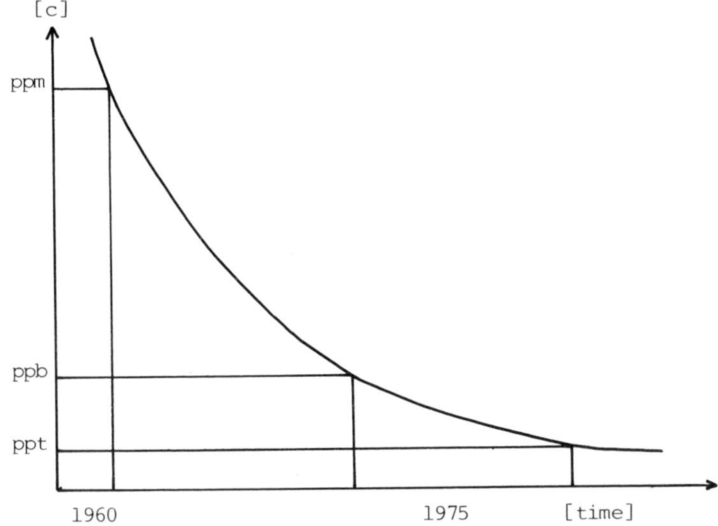

One can thus conclude that because of the increased efficacy of the available analytical methods, it will practically always be possible to detect residues, but these residues have very low concentrations and, above all, they are not necessarily toxic.

In view of the above it appears that the risks that residues represent for the consumer must be analyzed and one must give them their toxicologic dimension. This must be done by the scientific evaluation of their toxic potential.

3.   <u>Toxicological significance of residues</u>

Since residues only exist at very low doses in foods of animal origin, it is obvious that they present no risk of <u>acute</u> toxicity. Nobody has been intoxicated or will be intoxicated from eating foods containing residues once or many times in a non-repetitive manner in a lifetime. On the contrary one can imagine that the <u>regular intake</u> of small quantities of the <u>same</u> substance could at length lead to toxic symptoms : various insidious organic attacks, allergy and even cancer.

This is where the only possible risk lies. It has lead scientists, who are interested in these problems, to classify veterinary drugs into two large groups :
- those whose use is limited, that are applied to small numbers of animals at a time. They are called "occasional drugs". They do not present any risks for the consumer;
- those which are administered to large groups of animals, on a regular basis, as prophylactic treatments or in industrial farming : anabolic agents, antiparasite products, antibiotics, etc. These drugs called "mass" therapeutics must be strictly controlled and monitored in their distribution and usage as they are of concern for public health.

An important remark when talking about the toxicity of veterinary drug residues : there must be no confusion between hazards from drugs used at large doses in human therapy and risks from residues of the same substances which exist in low concentrations in the food. An obvious example is, in my opinion, that of diethylstilbestrol (DES). DES has been used legally and is still used illegally in a number of countries as anabolic agent. DES proved to be carcinogenic in man. It had caused vaginal cancer in girls born from mothers treated during their pregnancy. However, if one compares the doses involved in this type of accident with the quantities found in meat from treated animals the following is observed :

Carcinogenicity of DES residues in the woman	
Epidemiologic studies (Meselson et al. 1975)	Total dose : 0.5 to 300 mg/kg 0.2 % vagin and/or cervix cancer
Minimum dose in the woman : 0.5 mg/kg, i.e. 25 mg for a 50 kg woman = 25,000 µg	Maximum dose in calf livers : 1 ppb = 1 µg/kg
Mass of liver to be ingested 25,000 : 1 = 25,000 kg 750 G OF LIVER DAILY FOR 100 YEARS	

An epidemiological study performed in the United States indicated a frequency of 0.2 % of vaginal cancer in girls born from treated mothers. The cumulative doses used ranged from 0.50 to 300 mg/kg, which is incredibly high. Based on the lowest dose, i.e. 0.50 mg/kg, and the levels of residues observed in implanted animals, it can be estimated that a woman would have to eat 750 g of liver daily during 100 years to reach that dose. This is certainly a reassuring observation. In view of its toxicological profile, its important oral activity due to its lack of metabolization and its carcinogenic potential, the prohibition of DES remains nevertheless a good thing as long as this prohibition can effectively be controlled.

It can be said in general that there are potential risks for a consumer who eats food from animals treated with veterinary drugs. The problem is now to evaluate this risk, or in other words, to give residues their toxicological significance.

Through various toxicological studies carried out on laboratory animals (rat, mouse, dog, hamster, etc...) and in particular through long term studies involving a regular intake of the product, it is possible to define for man an acceptable daily intake (ADI). This ADI expressed in mg/kg of live weight is obtained by the following calculation : the no-effect level dose noted on the most sensitive species is divided by a safety factor ranging from 100 to 1,000 depending on the nature of the effects observed. This safety factor takes into account the possibility of a larger sensitivity of the human species compared to the animal species tested. This is essential to calculate and understand the notions of tolerance which we will explain below. There are three types of tolerance : toxicological tolerance, practical tolerance and analytical tolerance.

The <u>toxicological tolerance</u> is derived from the ADI. When multiplying the ADI by the average weight of man (70 kg) one obtains the daily quantity that this individual can eat without risk. When dividing this by the daily food intake, for example 500 g of meat, one obtains the maximum concentration that can be accepted in this meat. Below this concentration or toxicological tolerance, the consumer's health is of course protected.

However, one usually observes that when the veterinary drug is used in the right conditions (dosage, administration route, appropriate withdrawal time), the residue levels are below the toxicological tolerance. One thus choses, from the regulatory point of view, to retain this <u>practical tolerance</u> which actually indicates a correct use of the product. By doing this, an additional safety factor is brought.

The practical tolerance must obviously be controlled. Therefore, an analytical method with sensitivity below the practical tolerance must be available. This sensitivity limit is called <u>analytical tolerance.</u>

In summary, if veterinary drugs can leave residues in food from treated animals, it must be emphasized that these residues are always quantitatively very low and that their presence is not necessarily toxic. We have means of evaluating their toxicological significance and thus to define tolerances which are required to ensure the protection of consumers' health. These tolerances must obviously be respected and this implies a number of conditions to be filled.

The available drugs must firstly be quality products. This means that producers must manufacture them correctly and perform the necessary trials to check their efficacy, their usefulness and their toxicity. Approval to market the products will depend on this. The drugs must be administered in normal conditions of use (dosage, administration route, withdrawal times, indications must be respected). This is the responsibility of the veterinarian and the farmer. Efficacious controls must of course be performed and the authorities should be able to achieve this. Finally, and this is fundamental, consumers must be perfectly and honestly informed and even educated, since badly understood risks lead to uncontrolled fears. This is the only way to avoid conflicts where passion and irrational behaviour oppose to science. Such information is commonly provided in the United States and it is abnormal that it does not develop similarly in Europe.

In any case, it should not be forgotten that to live means to risk one's life and to conclude I shall quote Judge Warren Burger of the US Supreme Court (1980) who said : "Perfect safety is a chimera, regulation must not strangle human activity in a search for the impossible". I fully concur with this view.

RESUME

SANTE PUBLIQUE ET MEDICAMENT VETERINAIRE

André Rico,
Laboratoire de Toxicologie Biochimique et Métabolique (INRA)
Ecole Nationale Vétérinaire de Toulouse,
F-31076 Toulouse Cedex, France.

Se nourrir, c'est prendre un risque. Il existe dans l'alimentation de nombreux "composés naturels toxiques" et ces produits sont peu étudiés. Certains sont connus, d'autres non, et ils peuvent en fait être très agressifs (Aflatoxine). Dans ces conditions, opposer au plan toxicologique les composés naturels et de synthèse apparaît comme un non-sens scientifique.

L'utilisation des médicaments vétérinaires procure un certain bénéfice, mais il existe naturellement une probabilité de résidus. Il y a dix ans, la notion simple de tolérance 0 était admise. Le développement des techniques analytiques l'a fait rejeter. Il y a toujours un résidu, mais il est toujours quantitativement faible et n'est pas forcément toxique. Au plan de la toxicité, il ne faut d'ailleurs pas confondre celle des résidus avec celle des médicaments humains ou vétérinaires : les doses d'ingestion ou d'administration en sont très différentes.

Il faut donner au résidu sa dimension toxicologique au travers de l'appréciation scientifique de sa toxicité : c'est sa signification toxicologique. Les dangers doivent être évalués en particulier par des études de toxicité chronique, où la répétition de la prise est importante (notion d'ingestion régulière). A partir de cette analyse, il est alors possible de définir des tolérances indispensables pour assurer la protection de la santé des consommateurs, ces tolérances doivent naturellement être respectées. Elles ne pourront l'être que par l'utilisation correcte et raisonnée des médicaments de qualité et par des contrôles adéquats.

La sécurité parfaite est une chimère : vivre c'est risquer sa vie. Bien se nourrir, c'est aussi disposer d'une alimentation variée qui diminue en fait la probabilité d'ingestion régulière d'un même composé naturel ou artificiel.

ZUSAMMENFASSUNG

DAS GESAMTPROBLEM DER SICHERHEIT

André Rico,

Laboratoire de Toxicologie Biochimique et Métabolique (INRA),

Ecole Nationale Vétérinaire de Toulouse,

F-31076 Toulouse Cedex, Frankreich.

Sich ernähren heißt ein Risiko eingehen. Es gibt in unserer Nahrung zahlreiche natürliche Stoffe mit toxischer Wirkung, und diese Stoffe sind noch wenig untersucht. Einige sind bekannt, andere noch nicht, und es können auch sehr aggressive darunter sein, z.B. die Aflatoxine. Unter diesen Umständen erscheint als wissenschaftlicher Unsinn, auf toxikologischer Ebene natürliche und synthetische Verbindungen einander gegenüberzustellen.

Der Gebrauch von Tierarzneimitteln ist zwar von heilsamer Wirkung, doch besteht natürlich die Wahrscheinlichkeit, daß Rückstände bleiben. Vor zehn Jahren galt noch der einfache Begriff der Null-Toleranz. Durch die Entwicklung der analytischen Technik sind wir davon abgekommen. Es ist immer ein Rückstand vorhanden, allerdings stets in geringsten Mengen und nicht notwendigerweise toxisch. Auf toxikologischer Ebene sollte man nicht Toxizität der Rückstände mit der Toxizität human- oder veterinärmedizinischer Arzneimittel verwechseln; dazu sind die Dosierungen zu verschieden.

Man sollte die toxikologische Dimension der Rückstände durch die wissenschaftliche Bewertung ihrer Toxizität betrachten : das ist ihre wahre toxikologische Bedeutung. Die Gefahren sind insbesondere durch Studien zur chronischen Toxizität zu erforschen, bei denen die Stoffe wiederholt aufgenommen werden. Ausgehend von dieser Analyse wird es möglich sein, die Toleranzbereiche zu definieren, die für den Gesundheitsschutz des Verbrauchers unumgänglich sind, und diese Toleranzen müssen natürlich eingehalten werden. Das kann nur durch die korrekte und vernünftige Handhabung qualitativ hochstehender Medikamente und durch entsprechende Kontrollen geschehen.

Die vollkommene Sicherheit ist ein Trugbild : Leben heißt das Leben zu riskieren. Eine gesunde Ernährung sorgt auch für Abwechslung, und diese verringert die Wahrscheinlichkeit, daß die gleiche natürliche oder synthetische Verbindung regelmäßig aufgenommen wird.

VETERINARY DRUGS AND HUMAN HEALTH
SPECIFIC ASPECTS OF SAFETY

Michiel DEBACKERE,
Institute of Pharmacology and Toxicology,
Faculty of Veterinary Medicine,
State University of Ghent (Belgium)

Any new drug used in animal as well as in human health, before it becomes available on prescription for treatment, has to be submitted to a set of tests and the history of its research and development can be divided broadly into three parts. First the desired and finally proposed pharmacological activity has to be discovered and proved in an appropriate laboratory testing system. Secondly, the compound has to be subjected to detailed toxicological studies? and the judgement made that it can be administered to men and animals without risk of serious hazards. Thirdly, efficacy and reasonable safety have to be confirmed in clinical trials in patients.

For veterinary drugs, however, in contrast to drugs proposed for human medicine, a supplemental fourth part is required in the development. Since the products of a lot of treated animals such as meat, milk and eggs destined for human consumption can still be contaminated at the time of consumption with residues of the therapeuticals used, consumers are indeed asking for guarantees of safety. Therefore, additional assurances of safety are required from the pharmaceutical companies who must carry out further studies on the nature and quantity of residues and their possible toxicity in experimental animals.

However, it must be agreed that there is no drug in existence which is without toxicity and that a drug which has no toxicity will never be discovered perhaps. There are no absolutes in drug safety : the use of potent, biologically active substances can rarely, if ever, be free from hazard, and so long as drugs are used adverse reactions are liable to occur. Irrational belief in the value of inactive medicines and the equally irrational fear of harmful consequences from useful drugs are two obstacles encountered nowadays. Therefore one will have to succeed in bringing about a fundamental change in the public attitude towards drugs. The public is so frequently misinformed by the news media. Real or supposed side-effects are featured in an often sensational and frightening manner. It would be much better if these opinion-making media were to be used to communicate objective information on drugs

which would lead to their judicious use. In contrast to some rash allegations
formulated by certain so-called social critics accusing pharmaceutical industries of only wanting production and profit, it can be accepted that each
self-respecting pharmaceutical company, just as the clinician, is interested
in the development of methods which make it possible to predict as much as
possible the eventual toxicity of new compounds that are to be marketed.

The organisation of this symposium under the topic "Wholesome food for
all - Manger mieux et à sa faim" is a proof of this attitude.

In the preceding session of this symposium it was demonstrated that
the consumer needs enough and good quality, i.e. safe, food and that there is
an absolute need for animal health products within the frame of modern ecolo-
gical and hygienic conditions of farming. Between these two apparently con-
flicting positions, on the one hand the customer's demand and on the other
hand the farmer's needs, and within the frame of regulations and legislation,
it is the moral obligation and responsibility of the pharmaceutical industry
to develop suitable and safe products. It is indeed essential that these animal health products do not endanger human health and that their use gives us
as much guarantee as possible of safety.

Hence the second session of this symposium : the problem of safety.
As has been demonstrated by the first speaker in this session in the first
instance there is a general problem of safety. "To take food always means to
take risks", as there are always natural, partly known, partly unknown, toxic
substances present in our food. But the use of veterinary drugs provides also
a supplemental possibility of contamination with residues of these artificial
substances. However, the toxicological significance of these residues must be
situated in the frame of the scientific approach of toxicity.

Toxicity is a multivalent property and is often considered as being of
two distinct types : acute and chronic toxicity. Because acute toxicity is
practically excluded in the case of residues, it is the chronic toxicity which
provides many of the problems of safety evaluation, particularly for induced
changes in cell behaviour, such as carcinogenesis, mutagenesis or immunolo-
gical disorders and for cell death, when cell regeneration may not be possible
as in some central nervous effects and teratogenesis.

Concerning the pharmacological and therapeutic effects of a new drug,
well tried pharmacological models now available frequently provide a reliable
indication of its qualitative effects in man and, in some instances, a useful
guide to its potency. Reliable indications of potential toxicity in man,
however, are less easy to obtain. Difficulties arise because toxicological
studies demand experiments upon whole animals - models that are exceedingly

complex. In the laboratory investigations concerned with a new drug, a pharmaceutical company exercises its own control over the methods it will use to evaluate efficacy, but to a very large extent it must follow a prescribed pattern in the evaluation of toxicity. Toxicological testing methods and systems have been set up partly by scientific experience and consensus of opinion between toxicologists, but also partly as a result of special demands of some regulatory authorities, which are not always free from certain political influences as the opinion of misinformed and incompetent non-experts is frequently heard. Although these non-experts are asking for absolute safety with absolute certainty, the expert can only give an assurance of relative or probable safety. To give this assurance of probably safety, the expert has to perform a lot of laboratory studies and toxicological tests.

A first category of tests consists of those that are designed to evaluate in detail specific types of toxicity, as the intended use of a compound may require that an estimation of the order of safety from certain specific toxicities should be investigated. In the following papers four particular issues will be reviewed, covering the possible significant toxicity aspects of drugs used in animal health products. Three of the selected subjects are more or less specific potential problems of particular classes of drugs. The fourth subject, allergy, would apply to every class of products.

In a first paper, Professor Delatour will treat the problem of embryotoxicity and teratogenesis for anthelmintics. Chemicals that adversely influence the development of the foetus produce effects that vary from lethal to malformation and retardation of growth. Collectively these responses are termed embryotoxic effects. Some agents are predominantly lethal whereas others are predominantly able to produce malformation of the foetus. Because the malformation of the foetus is called terata, the agent that produces terata is termed a teratogenic agent and the pathogenesis of malformation is called teratogenesis. In order to evaluate the risks of residues from teratogenic anthelmintics in food of animal origin for the consumer, a description will be given of the relation between known metabolites and embryotoxicity.

A second important group of drugs used in animal production, partly legally, partly illegally, is anabolic, mostly hormono-mimetic agents. Since there is a heavy opposition from consumers and their organisations, especially in Europe, against their use because they are claimed to be carcinogenic, this forum offers us an ideal opportunity to make an evaluation of these substances for their suspected mutagenic and carcinogenic potential, another kind of specific toxicity. Mutagenesis is the induction of those alterations in the genetic information content (DNA) of an organism or cell that are not due to the normal process of recombination. This genetic damage can occur in both somatic and germinal cell lines. Somatic mutations in an adult may lead to cancer. Genetic alterations in germ cells, on the other hand, are passed on to further generations. With present knowledge a cancer cell must be regarded as a mutated normal cell, i.e. a cell sufficiently like to normal to survive in the environment provided by the body and sufficiently unlike it to disobey the normal rules of tissue growth. Theoretically, cancer may arise in several ways. This will be illustrated in Dr. Roe's paper.

As antibiotics and related anti-infectious chemotherapeutics form by far the largest group of drugs used in veterinary medicine, it is a matter of fact that during this symposium the problem of their safety will also be treated. For these substances safety is determined not only by their potential residues in food of animal origin, but far more by the problem of resistance development in some important pathogenic strains of bacteria, which will have important consequences for the treatment and control of infectious diseases in animals as well as in men. These problems will be treated for animals and for men respectively by Dr. Taylor and Dr. Tancrède.

A further kind of specific toxicity related to the use of animal health products and the risks of their residues for the consumers concerns hypersensitivity or allergic reactions. Although these reactions can be provoked by all classes of products as they are substances strange to the body, antibiotics are the class of drugs mostly involved. Residues of veterinary drugs are frequently incriminated as responsible for intolerances of food origin observed in human beings. In first instance we can ask what the nature and the mechanism of this risk are and how this risk can be evaluated by means of specific toxicological tests. A second aspect is the dose-effect relationship and the determination of a no-effect level for allergenicity by veterinary drug residues. All these questions will be treated in the paper presented by Mrs. Burgat-Sacaze.

Having examined the specific safety aspects of the three major classes of veterinary products : anthelmintics, anabolic agents and antibacterials, it can be agreed that this covers the bulk of animal health products used. This means quantitatively and qualitatively more than 90 % of the products used in veterinary medicine.

But the pharmaceutical industry should be aware, and it is, I believe, that besides these four specific aspects of the study of drug toxicity, there are still other kinds of toxicity which can be provoked by animal health products and which have to be studied with the same care and anxiety in the interest of human health, but with a different approach. Indeed, if drug safety were tested only on the basis of the foregoing tests, some side-effects and toxic reactions could be overlooked with the consequences that certain dangerous drugs would be admitted for prescription.

In first instance acute and still more prolonged and chronic toxicity tests have to be continued despite their shortcomings. As the former is giving us only information on the order of the lethality of the compound, the objective for the other two procedures is generally to evaluate and characterize all effects of compounds, when they are administered repeatedly on a daily basis over a periods of several months or years. There is considerable evidence indicating that these tests will reveal most forms of toxicity with the exception of the specific toxicities just mentioned. These toxicological procedures are mostly issued in writing by governmental regulatory agencies and must be performed as directed, so that toxicologists fear - not without reason - that any significant deviation from the generally accepted guidelines may be questioned by reviewers at regulatory agencies. As a consequence, in order to avoid any delay, the guideline which will satisfy the most demanding health authority becomes the generally accepted procedure. Nevertheless a toxicological investigation may no longer be considered as a mere classification of undesirable morphological, functional or biochemical effects observed after the administration of drugs. Such a study must also be concerned with the basis mechanisms of drug action which underlie the observed toxic effects. Knowledge of the pharmacokinetics and the biochemical transformation of a drug are important in understanding drug action. Such studies have been routine in pharmacological research for some time and it is now mandatory that they should be a routine part of all toxicological studies. It is now possible to obtain the necessary pharmacokinetic data for most drugs. Pharmacokinetic data do not only mean measurements of blood levels, calculation of half-live periods, studies of excretion and route of excretion, but also the knowledge of the pattern of metabolism. In the case of drug metabolites, it is becoming evident that toxicity may be related not to terminal metabolites, but to the formation of short-lived metabolites such as epoxides, hydroxylamines and nitroso-compounds or anionic radicals. The best proof for the value of this knowledge is the study of chloramphenicol toxicity. This is characterized by an irreversible and dramatic aplastic anaemia. Nowadays metabolic studies of chloramphenicol have revealed that this is probably provoked by the one-electron nitro radical anion derivative and not by a covalent binding of the dechlorinated dichloroacetamide residue.

Since the enzymes involved in metabolic transformation of drugs exist according to the genetic template characteristics of each member of a population of organisms, genetic defects in members of a species may result in a deficiency or complete lack of certain enzymes. Such genetic defects within members of a species have been shown to be responsible for some types of toxicity from drugs. Therefore, the study of genetic factors that influence toxi-

102

city can be of great value.

It should also be emphasized that toxic effects are sometimes related to the simultaneous use of several drugs. Therefore possible drug interactions should be investigated experimentally as a normal part of toxicological studies. Here again, pharmacokinetics and studies of drug metabolism do contribute to the understanding of the mechanisms involved in drug interaction. This interaction can take place at the level of intestinal absorption, renal excretion, protein binding, etc., but may also be a result of the inhibition or stimulation of liver microsomal enzymes involved in the metabolism of a number of drugs. A lot of such interactions between two drugs are very well known nowadays.

Another aspect of residue toxicity that ought to be incorporated in each test procedure is the possible binding to proteins and other macromolecules and the character of this binding, as protein binding can have a lot of toxicological implications and binding on macromolecules can be a step in potential mutagenicity and carcinogenicity. An example for this toxicological aspect is 17-β-trenbolone. A large proportion of unidentified residues appears to be irreversibly bound to macromolecules. As their bioavailability is low, they do not present a potential hazard to the consumer. However, the binding of free residues to macromolecules in the consumer may be of more importance.

In order to determine whether any adverse effect would be produced in man by consuming meat from treated animals, a relay or refeeding study can be undertaken. This study involves feeding various tissues from treated target species, e.g. beef cattle, to experimental animals such as rats and mice. These tissues may contain metabolites to which man can be exposed. This particular method has been used to demontrate the relay toxicity of DES.

Finally, it should not be forgotten that part of administered animal health products leaves the animal body partly metabolised and partly unchanged with the excreta and reaches the ecosystem. Therefore, more and more attention should be paid to the influence of these products on this system and thus ecotoxicity should be previewed and incorporated also in the toxicological evaluation scheme of animal health products, more especially for feed additives.

In this introduction I have tried to summarize some problems we are faced with during drug safety studies for animal health products especially in relation to the risks of residues for the consumer of food of animal origin. I know these are just a few examples of the problems we hope to discuss today. It will obviously be impossible to cover all the aspects of interest and thus it is desirable that an open discussion should complement the programme.

<u>References</u>

Davey, D.G., 1973. The cost of a drug. In : Toxicity review and prospect.
  V. Toxicology and the future of drug therapy. Proc. of the European
  Society for the Study of Drug Toxicity, Vol. XIV. Ed. Excerpta
  Medica/American Elseviers, pp. 187-193.
Debackere, M., 1983. Environmental pollution : the animal as source indicator
  and transmitter. Proc. from the 2nd E.A.V.P.T. Congress, Toulouse,
  Veterinary Pharmacology and Toxicology, MTP Press Limited, pp. 595-608.
Delatour, P., 1983. Evaluation of drug residues in animal tissues. Proc.
  from the 2nd E.A.V.P.T. Congress, Toulouse, Veterinary Pharmacology and
  Toxicology, MTP Press Limited, pp. 659-670.
Dunne, J.F., and Manser-Jones, 1973. Review and preview in safety assessment.
  In : Toxicity review and prospect. V. Toxicology and the future of drug
  therapy. Proc. of the European Society for the Study of Drug Toxicity,
  Vol. XIV. Ed. Excerpta Medica/American Elseviers, pp. 194-198.
Garattini, S., 1973. Pharmacokinetics and Toxicology. In : Toxicity
  review and prospect. Proc. of the European Society for the Study of Drug
  Toxicity. Vol. XIV, Excerpta Medica/American Elserviers, pp. 7-9.
Janssen, P.A.J., 1973. The prospects of drug therapy. In : Toxicity review
  and prospect. V. Toxicology and the future of drug therapy. Proc. of the
  European Society for the Study of Drug Toxicity, Vol. XIV. Ed. Excerpta
  Medica/American Elseviers, pp. 181-183.
Loomis, G., 1978. Essentials of Toxicology. Lea and Febiger, Philadelphia.
Schmid, A., 1983. Chloramphenicolrückstände in Lebensmitteln tierischer Her-
  kunft als potentielle Ursache der aplastischen Anämie der Menschen. Dtsch.
  Tierärztl. Wchschr., 90, 6, 201-248.
Zbinden, G., 1973. Progress in Toxicology. Special Topies, Vol. 1. Springer
  Verlag, New York Inc.

RESUME

PRODUITS VETERINAIRES ET SANTE PUBLIQUE
ASPECTS PARTICULIERS DE SECURITE

Michiel DEBACKERE
Institut de Pharmacologie et Toxicologie
Faculté de Médecine Vétérinaire
Université de Gent (Belgique)

Tout nouveau produit, qu'il soit destiné à la médecine vétérinaire ou à la médecine humaine, doit subir une série de tests avant de pouvoir être utilisé. Son développement peut être divisé en trois phases : premièrement, celle de la découverte et de la preuve de l'efficacité pharmacologique par des méthodes de laboratoire appropriées; deuxièmement, celle de l'étude toxicologique détaillée qui déterminera si le produit peut être administré aux animaux ou à l'homme sans danger; et troisièmement, la confirmation de l'efficacité et de l'innocuité du produit par des tests cliniques sur des patients.

Pour les produits vétérinaires, une quatrième phase s'impose néanmoins : il est nécessaire d'évaluer les risques que pourrait présenter pour le consommateur la contamination de denrées d'origine animale par les résidus de médicaments administrés pendant la vie de l'animal. Il est indispensable, en effet, que les produits utilisés en santé animale ne menacent pas la santé publique et que leur utilisation offre un maximum de garanties de sécurité.

La toxicité est une notion polyvalente et l'on en distingue généralement deux types : la toxicité aigüe et la toxicité chronique. La première peut pratiquement être exclue dans le cas de résidus, et c'est surtout la seconde qui cause le plus de problèmes d'évaluation, particulièrement lorsque des modifications interviennent au niveau du comportement cellulaire, comme c'est le cas pour la carcinogénèse, la mutagénèse, lors de désordres immunologiques ou lors de la destruction irréversible de cellules, comme dans certains effets neurologiques ou tératologiques.

Les tests de tératogénèse et d'embryotoxicité, les tests de mutagénèse et de carcinogénèse, les essais bactériologiques et les tests d'allergie, développés dans la seconde session de ce symposium couvrent quelques 90 % des risques toxiques des médicaments. D'autres formes de toxicité existent et font également l'objet d'études toxicologiques. Elles peuvent être causées par des métabolites intermédiaires ou des interactions entre plusieurs médicaments. Les facteurs génétiques des enzymes qui interviennent dans la métabolisation ou la liaison des résidus peuvent aussi avoir leur importance. Enfin comme les produits sont partiellement excrétés soit inchangés soit sous forme de métabolites, leur effet sur l'écosystème devrait aussi faire partie de leur évaluation toxicologique.

ZUSAMMENFASSUNG

VETERINÄRERZEUGNISSE UND ÖFFENTLICHES GESUNDHEITSWESEN

BESONDERE SICHERHEITSTECHNISCHE GESICHTSPUNKTE

Michiel Debackere

Institut für Pharmakologie und Toxikologie

Veterinärmedizinische Fakultät

Universität Gent, Belgien

Jedes neue Erzeugnis, ob es für die Veterinär- oder für die Humanmedizin bestimmt ist, muß vor seiner Ingebrauchnahme einer Reihe von Prüfungen unterzogen werden. Seine Entwicklung läßt sich in drei Phasen unterteilen : erstens, die Phase der Entdeckung und des Nachweises der pharmakologischen Wirksamkeit mit Hilfe angemessener Labormethoden; zweitens, die Phase der ausführlichen toxicologischen Untersuchung, aus der hervorgeht, ob man Tieren oder Menschen das Erzeugnis gefahrlos verabreichen darf; und drittens, die Phase der Bestätigung der Wirksamkeit und der Harmlosigkeit des Erzeugnisses mit Hilfe von klinischen Prüfungen an Patienten.

Bei Veterinärerzeugnissen ist jedoch eine vierte Phase unbedingt notwendig : es sind nämlich die Risiken zu beurteilen, die sich für den Verbraucher infolge der Kontamination von Nahrungsmitteln tierischen Ursprungs durch die Rückstände von Arzneimitteln ergeben können, die während der Lebensdauer des Tiers verabreicht worden sind. Es ist, in der Tat, unbedingt erforderlich daß die auf dem Gebiet der tierischen Gesundheit verwendete Produkte die Volksgesundheit nicht gefährden und maximale Sicherheit hinsichtlich ihres Einsatzes gewährleistet wird.

Die Giftigkeit ist ein polyvalenter Begriff und man unterscheidet im allgemeinen zwischen zwei Arten : akute Giftigkeit und chronische Giftigkeit. Bei den Rückstandsfällen läßt sich die akute Giftigkeit praktisch ausschließen und es ist besonders die chronische Giftigkeit, die die meisten Problemen bei der Beurteilung verursacht, insbesondere wenn Änderungen des Zellenverhaltens auftreten wie bei der Carcinogenität, der Mutagenität, bei immunologischen Störungen oder bei der irreversiblen Zerstörung der Zellen, z.B. bei gewissen neurologischen und teratologischen Auswirkungen.

Die auf der zweiten Sitzungsperiode dieser Tagung geschilderten Mißbildungs- und Embryotoxizitätstests, die Mutagenitäts- und Carcinogenitätstests, die bakteriologischen Prüfungen und die Allergietests erstrecken sich auf etwa 90 % der toxischen Risiken der Arzneimittel. Es gibt weitere Formen der Toxizität, die ebenfalls toxikologischen Untersuchungen unterzogen werden. Sie können von intermediären Metaboliten oder den Wechselwirkungen mehrerer Arzneimittel verursacht sein. Die genetischen Faktoren der Enzyme, die zur Metabolisierung oder Bindung der Rückstände beitragen, können ebenfalls ihre

Bedeutung haben. Schließlich und weil die Produkte teilweise unverändert oder in der Form von Metaboliten ausgeschieden werden, sollte ihre Auswirkung auf das Ökosystem ebenfalls Bestandteil ihrer toxikologischen Beurteilung sein.

EMBRYOTOXICITY AND TERATOGENESIS :
CONSEQUENCES FOR VETERINARY DRUGS

Paul DELATOUR,
Laboratoire de Biochimie,
Ecole Nationale Vétérinaire de Lyon,
F-69752 Charbonnières Cedex, France

Introduction

Teratogenesis is due to the possible side effects of drugs. This special form of toxicity involves two types of consequences : the possible existence of a safety margin for the target animal in gestational circumstances and the possible risk related to the residues of these drugs in animal products (meat, milk, eggs) destined to human food.

We would like first to recall the basic law of teratogenesis, then to list the main chemicals responsible for teratogenic effects in humans and animals, and finally to analyse some information concerning the metabolism of these drugs. These documents will then help us to discern a general line of thought about the toxicological evaluation of residues of teratogenic drugs.

1.   Basic laws of teratogenesis

According to experimental data, it would appear that the general laws of embryotoxic induction defined for laboratory animals are also applicable to domestic animals. The main aspects useful to us are (a) species sensitivity or refractoriness, (b) low amplitude of teratogenic stimulus, (c) concept of no-effect level and (d) concept of a critical period.

a) Species sensitivity

The comparison of experimental results for a given teratogen in different animals points out that some species are sensitive while others are refractory. Parbendazole is teratogenic in sheep (1, 2, 3, 4, 5), but not in cattle (6) and pigs (7); mebendazole is a teratogen for rats (8) and not for sheep (9); oxfendazole is teratogenic for sheep (10) but not for cattle (11). Carbendazim is teratogenic for rats, mice, rabbits and hamsters (8, 12). Thalidomide is teratogenic for men, monkeys, rabbits and mice, but generally not for rats.

b) <u>Low amplitude of the teratogenic stimulus</u>

In connection with the different degrees of embryotoxicity of the teratogens, there exists a wide range of situations. For example, retinol, salicylate and griseofulvin are teratogenic only at very high levels, possibly not too far from toxic doses for the adult. On the contrary, in the most typical cases - corresponding to the highly active compounds - the teratogenic doses are very inferior to the toxic levels for the mother : parbendazole is teratogenic for sheep at 40 mg/kg whereas the maximum tolerated dosage by adults is 1000 mg/kg; the corresponding figures for thalidomide in dogs are 30 and 1500 mg/kg respectively (13, 14). In addition, there is often a narrow margin between the optimum teratogenic dosage and that which is embryolethal : in dogs, the incidence of thalidomide-induced malformations (expressed in percentage) increases according to the following doses : 500 (15), 400 (16), 200-100 (17) and 30 mg/kg (13). At high dosages, the embryolethal effect may limit the teratogenic effect by provoking the death of the most sensitive embryos.

c) <u>Concept of no-effect levels</u>

In every type of teratogenic compound, it is always possible to determine experimentally a dose which induces neither embryolethal nor teratogenic effects. This is true for both laboratory and target animals.

The following figures point out that in the Sprague-Dawley rat, the no-effect level of albendazole administered by gastric tube from day 8 to day 15 of pregnancy is 6.0 mg/kg (18).

TABLE 1

Dose response of albendazole (ABZ) in the rat

	Embryo-lethality (%)	Teratogenicity : external (%)	skelettal (%)	Foetogenicity : bodyweight (g)
Control	7.1	0.2	1.4	3.90
ABZ = 5.30 mg/kg	6.0	0.3	1.7	3.89
6.00	4.6	0.0	1.8	3.87
6.62	10.9	0.0	3.2	3.46
8.83	18.4	14.4	39.6	3.13
10.60	80.2	94.7	100.0	2.03
13.25	100.0	–	–	–

Similar figures have been published about numerous embryotoxic substances for rats and mice. Identical conclusions can be drawn for example in

sheep when the maximum no-embryotoxicity levels are for albendazole, oxfendazole and parbendazole : 10, 15 and 30 mg/kg respectively on day 17 of pregnancy.

d) Concepts of critical period and single dose effect

Experimental studies have shown that every teratogenic substance can
produce morphological abnormalities only during organogenesis, a relatively
short period, generally limited to the first third of the pregnancy. During
the initial stages corresponding to the "refractory period", the embryo
expresses an all-or-nothing response to the aggression of the toxic drug :
either embryolethality or a "regulation process" will occur giving rise to
normal further development. In the latter stages of histogenesis of dys- or
aplastic type these are no longer possible and troubles of general growth or
tissular differentiation appear. Classically, the teratogenic period covers
the following : rabbits, from day 8 to day 16; cats, from day 5 to 20; dogs,
from day 8 to 25; pigs, from day 10 to 24; and sheep, from day 10 to 25.

Unlike laboratory animals such as rats and mice when the critical
period starts immediately after implantation, in other species, malformations
can be induced before this stage : for example in sheep, exencephaly is
possible on day 12 and cyclopia on day 14 (19), whereas implantation takes
place on day 15-16. Identical remarks have been made in cats with 10-methyl-
folic acid and dogs with thalidomide.

Finally, the reproductive organs and the central nervous system show a
long and delayed period of structural and histological maturation. For
instance, trichlorfon in pigs is responsible for an ataxia syndrome when
administered between day 45 and day 63 of pregnancy (20), whereas the
classical period ends on day 25.

2. Main teratogens in humans and animals

In the human being, a lot of epidemiological studies have shown the
possible teratogenicity of some therapeuticals. Few facts are well established at the present time, so that among the widely used drugs, less than 10
have been really proven to be teratogenic in the human embryo.

Some examples are : anti-folates such as aminopterin (21), previously
used as an abortifaciant; thalidomide (22, 77); the antiepileptics, barbitu-
rates (23, 24); phenytoin (25, 26, 27); trimethadione (28, 37) and recently
valproic acid (29, 30); progestins (31); diazepam (32); diethylstilbestrol
(DES) (33, 34, 35) and anticoagulants (36).

Finally a recent retinol-like compound, etretinate, very potent
against psoriasis is strictly prohibited during pregnancy, because it is sus-

pected to be a teratogen in man. Experimentally, all these compounds also developed teratogenic effects in laboratory animals.

Described as teratogens in cats are : 10-methyl-folic acid (37); thalidomide (38); griseofulvin (39, 40, 41); salicylic acid and methotrexate (42).

For practival reasons probably, few investigations have been performed on dogs. Even then, some papers are available in this area referring to : diazo-oxo-norleucine (43); thalidomide (13, 15, 16, 17); chlortetracycline (44), carbaryl (45), vitamin A (46).

In pigs the following chemicals have been proven experimentally to disturb embryo-foetal development : retinol (47), thalidomide (47), methallibure (48, 49, 50); metrifonate (20). Also clinical observations have established that ingestion of the plant Nicotiana tabacum (52) by sows while pregnant can cause birth defects in piglets.

In sheep, many benzimidazole anthelmintics are well known as typical examples of teratogens in ewes, at least experimentally. These are parbendazole (53, 4, 5); cambendazole (54), oxfendazole (10) and albendazole. On the contrary, mebendazole (9) and oxibendazole (55, 76) do not seem to induce morphological abnormalities in lambs. Also, numerous plant alcaloids are known as embryotoxic agents in sheep, such as Veratrum californicum (56), Astragalus lentiginosis (57); Lupinus cosentinii (58, 59). Aminopterin is also teratogenic (78).

If we ignore the case of a clinical suspicion concerning thio-uracyle (60), the literature provides no example of a drug capable of teratogenic effects in cattle, even among compounds that are teratogenic in sheep (11, 6). But different plants are known to disturb embryo-foetal development like Lupinus caudatus (61), Nicotina glauca and Astragalus lentiginosis (57).

3.   <u>Teratogenic hazards of drug residues in humans</u>

Residues of veterinary drugs or pesticides in animal products (meat, milk, eggs) are the result of animal metabolism. Classically we can distinguish different residual fractions among "total residues" (arbitrarily expressed in ppm of total radioactivity), according to methodological criteria :

<u>Intractable residues</u> :

a) incorporated radioactivity into endogenous metabolism

b) "hydrid molecules" (64)

c) hydrophilic compounds (conjugates)

d) covalently bound adducts on macromolecules.

<u>Extractible residues</u> :

e) parent compound

d) free metabolite.

Among intractable residues a-fraction is practially an artefact and its innocuousness is unanimously accepted, while b-fraction is a curiosity that has been described only in a limited number of examples (64, 65) and which is of no general interest. The c-fraction of conjugates represents, from the toxicological viewpoint, a mixture of inactive compounds destined to kidney elimination - in the most frequent cases - and of reactive chemical species that are able, in the target animal tissues, to be transformed into adducts. Finally, d-fraction is probably the most fascinating because of its frequency in all animal species (mammals, birds) and of its very long half-life.

These adducts are well documented as responsible for many toxic manifestations, including teratogenicity (66, 67) in animals as well as in man. In other circumstances, we are unable to correlate bound radioactivity with toxic effects (68, 69).

The situation is totally different in food hygiene when the adducts are ingested by a relay organism such as the consumer. Different methodologies have been developed for investigating the possible toxicity of intractable residues known as relay-toxicity, relay-bioavailability and target-impact-relay-assay. At present, none of these methods are sensitive or accurate enough to definitely guarantee the innocuousness of this residual fraction in relay circumstances. Yet, a survey of available published results in combination with theoretical considerations seems to argue in favour of their safety. For example, the main adduct of the carcinogen dimethylnitrosamine, 7-methylguanine, does not seem to be incorporated into nucleic acids of the liver, intestine and rat foetus (70), after oral administration, and does not present carcinogenic effects in the same animal species. Identical experiments have been performed with DNA-bound aflatoxin whose bioavailability according to theoretical calculations related to the sensitivity of the detection - is at least 4,000 times lower than that of free aflatoxin (71).

A totally different qualitative interpretation must be drawn concerning extractible residues of teratogenic drugs. In addition to the presence of unchanged teratogenic compounds in animal tissues and production (milk, eggs) extractible residues also include free metabolites : meclizine (72), chlorcyclizine (73). In the area of specific veterinary drugs, more or less exhaustive experimental studies of the metabolism-embryotoxicity relationship have shown that some compounds act per se (for example oxfendazole), other possess an identified teratogenic metabolite (for example benomyl, febantel). It must be noted that for some non-teratogenic compounds, such as fenbendazole or thiophanates, a teratogenic metabolite possibly present in animal products has been traced (74). For these reasons, a working group of the French

Pharmacopoeia developed a scheme of toxicological evaluation adapted to these problems (75). In short, this procedure consists of the following steps : identification of free metabolites present in edible tissues, synthesis of chemical samples, experimental screening of their individual teratogenic possibilities and determination of their no-effect level, discrimination between safe metabolites and those able to be metabolically converted into teratogenic compounds, study of the residual decrease versus time of the toxic and potentially toxic metabolites and determination of a withdrawal period by taking into account the Sum of Potentially Toxic Metabolites (SPTM), the lowest no-effect level and the safety coefficient.

According to this method, calculation of the withdrawal periods leads often to small figures and possibly, for milk, to none at all. On the contrary, in cases where in a single milking after treatment, the level of SPTM is a dose higher than the ADI, the use of such a drug is prohibited for dairy cattle, regardless of a possible withdrawal period.

In short, two main residual fractions are of toxicological concern :
- covalently bound adducts with a long half-life for which, strictly speaking, neither toxicity nor innocuousness can presently be established experimentally by direct methods;
- extractible residues, with generally a short half-life, including compounds whose responsibility for teratogenic effects can be proved by animal studies. These, at least quantitatively, represent a real or potential risk for the consumer.

In addition to the qualitative aspect, one must take into account the quantitative one. Animal teratological experiments and general laws of pharmacology show that in chemical-induced teratology, there is a no-effect level for every teratogen.

Since we have very little information about the relative sensitivity of human embryos compared to animal embryos, the ADI is calculated according to the most sensitive animal species. The lowest no-effect level is then divided by an appropriate safety coefficient : 1,000 is often the accepted value for teratogenic effects. On the other hand, experimental and previously discussed (63) thermodynamic arguments allow us to take into account - when the embryotoxicity-metabolism relationship has been established - the decrease of the sum of potentially toxic extractible metabolites whose biological effects are well established.

This procedure will be illustrated by the following two examples :
a) Example of mebendazole in meat :
- Mebendazole is teratogenic in rats (8) and to a lesser extent in mice; rabbits are refractory. In rats - the most sensitive species - the no-effect

level of mebendazole is 8 mg/kg/day. Also, the study of embryotoxicity of individual metabolites showed that another metabolite – the carbinol corresponding to the parent compound – is teratogenic as well (8). Its no-effect level is approximately 4.95 mg/kg.

– The safety coefficient for teratogenic drugs is 1,000, so the ADI for humans will be 4.95 µg/kg/day or 346 µg/day.

– In lambs, 10 days after an oral dose of 25 mg/kg mebendazole, the liver contains 3.1 ppm of total radioactivity. The extractible fraction represents 2 % of the total radioactivity. Inside this fraction, the SPTM (parent compound plus alcoholic metabolite) is only of 26 % according to Rico et al. studies, that is to say 0.016 ppm. The intake of 100 g liver provides 1.6 µg/ day to the consumer (70 kg) or 0.023 µg/kg/day.

– Finally, it appears that ingestion of 100 g of sheep liver 10 days after a treatment with mebendazole contains a residual level of SPTM 346/1.6 = 216 times lower than the ADI. In other words, the daily intake for human in these conditions will be 215,217 times lower than the no-effect level in the most sensitive experimental animal species. An equivalent dose (mg/kg) would be attained by man after ingestion of 21,656 kg liver in a single meal !

       b) Example of albendazole in milk :
The general reckoning is similar to that of mebendazole.

– The no-effect level of albendazole in rat is 6.0 mg/kg. In addition, among 10 metabolites, only one is also teratogenic – the sulfoxide corresponding to the parent drug – and its no-effect level is 6.0 mg/kg (18).

– The ADI for the SPTM (albendazole plus its sulfoxide) is, with a safety coefficient of 1,000, 6 µg/kg/day or 420 µg/day.

– After a single oral dosage of albendazole (7.5 mg/kg) to cows, the level of albendazole on the first milking is very low (less than 0.02 ppm) while for the sulfoxide it is approximately 0.8 ppm. In these circumstances, the daily ingestion of the equivalent of 800 ml milk provides 640 µg, which is higher than the ADI.

      In spite of a very fast decrease of the teratogenic metabolite (no residues after 48 hours), the withdrawal periods generally adopted over European countries are 3 to 5 days. Once again maximum safety margin is given to the consumer.

Conclusion

      Teratological manifestations are part of the side effects of numerous drugs and pesticides. The study of the laws of anomalies induction shows that some special aspects are of great concern to us : mainly the greater or lesser

116

sensitivity of animal species, the notion of no-embryotoxic level and of a
teratological calendar, and the role of the drug metabolism.

On the other hand, improvements of analytical procedures in the sepa-
ration of "total residues" into different residual fractions in combination
with the metabolism-teratogenicity relationship information show that teratological risks are not related equivocally to all residual fractions. In every
case, this discrimination must be performed experimentally.

Teratological risks for pregnant women through veterinary drug
residues must be neither neglected nor exaggerated; after 10 years' experience
in the area of teratology and metabolism, we can assume that this is not a
major risk.

Multidisciplinary discussions - especially in this sensitive area -
are always useful and of reciprocal benefit. Yet, biologists and analysts
claim responsibility for decisions about toxicological evaluation of residues.
It is always possible to ask for higher safety margins and longer withdrawal
periods; one must also keep in mind that reasonable attitudes are founded on
the famous risks-benefits balance. To ask for a really natural food is to
return to the Middle Ages. The Neanderthal man had no contact with chemical
residues, but probably with mycotoxins : he died at the age of 25 !

References

1. Shone, D.K., Philip, J.R. and Fricker, J.M., 1974. The effects of Methyl-
       5(6)-Butyl-2-Benzimidazole Carbamate (Parbendazole) on reproduction in
       sheep and other animals : teratological study in ewes in the Republic
       of South Africa. Cornell Vet., 64, (suppl. n° 4), 69-76.
2. Szabo, K.T., Miller, C.R., and Scott, G.C., 1974. The effects of Methyl-
       5(6)-Butyl-2-Benzimidazole Carbamate (Parbendazole) on reproduction in
       sheep and other animals : teratological study in ewes in the United
       States. Cornell Vet., 64, (suppl. n° 4), 41-55.
3. Lapras, M., Deschanel, J.P., Delatour, P., Castellu, M. and Lombard, M.
       1973. Accidents tératologiques chez le mouton après administration de
       Parbendazole. Bull. Soc. Sci. Vet. et Med. Comp. de Lyon, 75 : 53-61.
4. Lemon, P.G. and Hancock, N.A., 1974. The effects of Methyl-5(6)-Butyl-2-
       Benzimidazole Carbamate (Parbendazole) on reproduction in sheep and
       other animals : teratological study in ewes in the United Kingdom.
       Cornell Vet., 64 (suppl. n° 4), 77-84.
5. Middleton, M.D., Plant, J.W., Walker, C.E., Dixon, E.T. and Johns, D.R.,
       1974. The effects of Methyl-5(6)-Butyl-2-Benzimidazole Carbamate
       (Parbendazole) on reproductionin sheep and other animals : teratolo-
       gical studies in ewes in Australia. Cornell Vet., 64 (suppl. n° 4),
       56-68.
6. Miller, C.R., Szabo, K.T., Scott, G.C., 1974. The effects of Methyl-5(6)-
       Butyl-2-Benzimidazole Carbamate (Parbendazole) on reproduction in
       sheep and other animals : effect in pregnant cow. Cornell Vet. 64,
       (suppl. n° 4), 85-91.
7. Hancock, N.A., Poulter, D.A.L., 1974. The effects of Methyl-5(6)-Butyl-
       2-Benzimidazole Carbamate (Parbendazole) on reproduction in sheep and
       other animals : effect of administration to pregnant sow. Cornell
       Vet., 64, (suppl. n° 4), 92-96.
8. Delatour, P. and Richard, Y., 1976, Propriétés embryotoxiques et antimito-

tiques en série Benzimidazole. Thérapie, 31, 505-515.

9. O'Brien, J.J., 1971. Effect of Mebendazole on reporductive performance in sheep. Unpublished doc. n° V-703.

10. Delatour, P., Debroye, J., Lorgue, G., and Courtot, D., 1977. Embryotoxicité expérimentale de l'Oxfendazole chez le rat et le mouton. Rev. Med. Vet. 153, 639-645.

11. Piercy, D.W., Reynolds, J., and Brown, P.R.M., 1979. Reproductive safety studies on Oxfendazole in sheep and cattle. Br. Vet. J., 135, 405-410.

12. Minta, M., Biernacki, B., 1982. Embryotoxicity of carbendazim in hamsters, rats and rabbits. Bull. Vet. Inst. Pulawy, 25, 42-52.

13. Delatour, P., Dams, R. and Favre-Tissot, M., 1965. Thalidomide : embryopathies chez le chien. Thérapie, 20, 573-589.

14. Kunz, W., Keller, H., and Mücker, H., 1956. N-Phtalyl-glutaminsäure-imid. Experimentelle Untersuchungen an einem neuen synthetischen Produkt mit sedativen Eigenschaften. Arzneim. Forsch. 6, 426-439.

15. Roberjot, G., 1966. De l'utilisaton de la chienne en vue de la recherche du pouvoir tératogène d'un médicament. Comm. Symp. Internat. Lyon Anim. Lab. Lyon, 19-20 Sept., C.R. 221-226.

16. Riemschneider R. and Brockmeyer, K., 1963. Teratogene Wirksamkeit organischer Verbindung. III - Weitere Thalidomid-Versuche. Z. Naturforschg. 186, 584-585.

17. Weidman, W.H., Young, H.H., Zollman, P.E., 1963. The effect of thalidomide on the unborn puppy. Staff Meet. Mayo Clin. 38, 518-522.

18. Delatour, P., Parish, R.C., Gyurick, R.J., 1981. Albendazole : a comparison of relay embryotoxicity with embryotoxicity of individual metabolites. Ann. Rech. Vet., 12, 159-167.

19. Binns, W., Shupe, J.L., Keeler, R.F. and James, L.F., 1965. Chronologic evaluation of teratogenicity in sheep fed Veratrum californicum. J. Am. Vet. Med. Ass., 147, 839-842.

20. Knox, B., Askaa, J., Basse A., et al., 1978. Congenital ataxia and tremor with cerebellar hypoplasia in piglets born by sows treated with Neguvon vet. (metrifonate, trichlorfon) during pregnancy. Nord. Vet. Med., 30, 538è545.

21. Thiersch, J.B., 1952. Therapeutic abortion with folic acid antagonist 4-aminopteroylglutamic acid administered by oral route. Am. J. Obstet. Gynecol. 63, 1298è1304.

22. McBride, W.G., 1961. Thalidomide and congenital abnormalities. Lancet ii 1358.

23. Seip, M., 1976. Growth retardation, dysmorphie facies and minor malformations following massive exposure to phenobarbitone in uteri. Acta Paediatr. Scand., 65, 617-621.

24. Bethenod, M., Frederich, A., 1975. Les enfants de mères traitées par antiépileptiques. Pédiatrie, 30, 227-248.

25. Loughnan, P.M., Gold, H., Vance, J.C., 1973. Phenytoin teratogenicity in man. Lancet, 1, 70-72.

26. Hanson, J.W., Smith, D.W., 1975. The fetal hydantoin syndrome. J. Pediatr. 87, 285-290.

27. Monson, R.R., Rosenberg, L., Hartz, S.C., Shapiro, S., Heinonen, O.P. and Slone, D., 1973. Diphenylhydantoin and selected congenital malformations. N. Engl. J. med., 289, 1049-1052.

28. German, J., Ehler, K.H., Kowal, A., Degeorge, E.V., Angle, M.A., Passarge, E., 1970. Possible teratogenicity of trimethadione and paramethadione Lancet, 2, 261-261.

29. Robert, E., Guibaud, P., 1982. Maternal valproic acid and congenital neural tube defects. The Lancet, October 23, 937.

30. Bjerkedal, T., Czeided, A., Goujard, J., Kallen, B., Mastroiacova, P., Nevin, N., Oakley, G.Jr., Robert, E., 1982. Valproic acid and spina bifida. The Lancet, November 13, 1096.

31. Barnes, A.B., 1979. Maternal progestins as a possible cause of hypospadias. N. Engl. J. Med., 17, 988.

32. Safra, M.J. and Oakley, G.P., 1975. Association between cleft lip with or

without cleft palate and prenatal exposure to diazepam. Lancet, 2, 478-480.

33. Henderson, B.E., Benton, B., Cosgrove, M., Baptista, J., Aldrich, J., Townsend, D., Hart, W. and Mack, T.M., 1976. Urogenital tract abnormalities in sons of women treated with diethylstilbestrol. Pediatr., 58, 505-507.

34. Bibbo, M., Al-Nequeeb, M., Baccarini, L., Gill, W., Newton, M., Sleeper, K., Soneck, M., and Wied, G.I., 1975. Follow-up study of male and female offspring of DES-treated mothers - a preliminary report. Journal of Reproductive Medicine, 15, 29-32.

35. Kaufman, R.H., Binder, G.L., Gray, P.M., and Adam, E., 1977. Upper genital tract changes associated with exposure in utero to diethylstilbestrol. Am. J. of Obst. Gynecol., 128, 51-59.

36. Shaul, W.L. and Hall, J.G., 1977. Multiple congenital animalies associated with oral anticoagulants. Am. J. of Obst. Gynecol., 128, 51-59.

37a. Zackai, E.H., Mellman, W.J., Neiderer, B. and Hanson, J.W., 1975. Fetal trimethadione syndrome. J. of Pediatr., 87, 280-284.

37b. Tuchmann-Duplessis, H. and Lefebvres-Boisselot, J., 1957. Les effets tératogènes de l'acide 10-méthyl-folique chez la chatte. C.R. Soc. Biol., 151, 2005-2008.

38. Khera, K.S., 1975. Fetal cardiovascular and other defects induced by thalidomide in cats. Teratology, 11, 65-69.

39. Gillitch, A., 1972. Griseofulvin, a possible teratogen. Can. vet. J., 13, 244.

40. Klein, M.F. and Beale, J.R., 1972. Griseofulvin : a teratogenic study. Science, 175, 1483-1484.

41. Scott, F.W., de la Hunta, A., Schultz, R.D., Bistner, S.I., and Riis, R.C., 1975. Teratogenesis in cats associated with Griseofulvin therapy. Teratology, 11, 79, 86.

42. Khera, K.S., 1976. Teratogenicity studies with Methotrexate, Aminopterin and Acetylsalicylic acid in domestic cats. Teratology, 14, 21-27.

43. Friedman, M.H., 1957. The effect of D-Diazocetyl-L-serine (Azaserine) on the pregnancy of the dog. J. Am. Vet. Med. Ass., 130, 159-162.

44. Savini, E.C., Moulin, M.A., and Herrou, M.F.J., 1968. Effets tératogènes de l'Oxytétracycline. Thérapie, 23, 1247-1260.

45. Smalley, H.E., Curtis, J.M., Earl, F.L., 1968. Teratogenic action of carbaryl in beagle dogs. Toxicol. Appl. Pharmacol., 13, 392-403.

46. Wiersig, D.O. and Swenson, M.J., 1967. Teratogenicity of Vitamin A in the canine. Fed. Proc., 26, 486.

47. Palludan, B., 1965. Swine in teratological studies. Swine in biomedical research, 51-78.

48. Vente, J.P., Wrathall, A.E., Hebert, N. and Hoskin, B.D., 1972. Quantitative anatomical study of Methallibure-induced malformations in piglets. Research in Vet. Sci., 13, 169-173.

49. Barker, C.A.V., 1970. Antigestation and teratogenic effect of Aimax (Metallibure) in gilts. Can. Vet. J., 11, 39.

50. King, G.J., 1969. Deformities in piglets following administration of methallibure during specific stages of gestation. J. Reprod. Fert., 20, 551-5.

52. Crowe, M.W., Pike, H.T., 1972. Congenital arthrogryposis associated with ingestion of tobacco stalks by pregnant sows. J. Am. Vet. Med. Ass., 162, 453.

53. Saunders, L.Z., Shone, D.K., Philip, J.R., and Birhead, H.A., 1974. The effects of Methyl-5(6)-Butyl-2-Benzimidazole Carbamate (Parbendazole) on reproduction in sheep and other animals : malformation in newborn lambs. Cornell vet., 64 (suppl. n° 4), 7-41.

54. Delatour, P., Lorgue, G., Courtot, D., and Lapras, M., 1975. Embryotoxicité expérimentale du Cambendazole (M.K. 905) chez le mouton. Bull. Soc. Sci. Vet. et Med. Comp. Lyon, 77, 197-203.

55. Delatour, P., Lorgue, G., Courtot, D. and Lapras, M., 1976. Tolérance embryonnaire de l'Oxibendazole chez le rat et le mouton. Rev. Med.

Vet., 152, (7-8), 467-470.

56. Binns, W., Thacker, E.J., James, L.F., and Huffmann, W.T., 1959. A congenital cyclopian-type malformation in lambs. J. Am. Vet. Med. Ass., 134, 180-183.

57. James, L.F., Keeler, R.F., Binns, W., 1969. Sequence in the abortive and teratogenic effects of locoweed fed to sheep. Am. J. Vet. Res., 30, 377-380.

58. Hawkins, C.D., Skirrow, S.Z., Wyburn, R.S., and Mc Howell, J., 1983. Hemimelia and low marking percentage in a flock of Merino ewes and lambs. Austr. Vet. J., 60, 22-24.

59. Allen, J.G., Fenny, R.E., Buckman, P.G., Hunt, B.R. and Morcombe, P.W., 1983. Hemimelia in lambs. Austr. Vet. J., 60, 283-284.

60. Mammericks, M., 1967. A propos d'un cas d'achondroplasie chez un foetus de l'espèce bovine. Rec. Med. Vet., 143, 239-244.

61. Shupe, J.L., James, L.F., and Binns, W., 1967. Observations on crooked calf disease. J. Am. Vet. Med. Ass., 151, 191-197.

62. Shupe, J.L., James, L.F., 1983. Teratogenic plants. Vet. Hum. Toxicol. 25, 415-421.

63. Burgat, V., Delatour, P., Rico, A., 1981. Bound residues of veterinary drugs : bioavailability and toxicological implications. Ann. Rech. Vet., 13, 277-289.

64. Caldwell, J., Marsch, M.V., 1983. Interrelationships between xenobiotic metabolism and lipid biosynthesis. Biochem. Pharmacol., 32, 1667-1672.

65. Hiremath, C.B., Olsen, G., Rosenblum, C., 1971. Incorporation of 1-0-ethyl -threonine into chick muscle protein. Biochem., 10, 1096.

66. Gordon, G.B., Spielberg, S.P., Blake, D.A., et al., 1981. Thalidomide teratogenesis : evidence for a toxic arene oxide metabolite. Proc. Nat. Acad. Sci. USA, 78, 2545-2548.

67. Martz, F., Failinger, C., Blake, D., 1977. Phenytoin teratogenesis : correlation between embryopathic effect and covalent binding of putative arene oxide metabolite in gestational tissue. J. Pharmacol. exptl. Therap., 203, 231-239.

68. Jaglan, P.S., Gosline, R.E., Neff, A.W., 1976. Metabolic fate of p-toluoylchloride phenylhydrazone (TCPH) in sheep. The nature of bound residues. J. Vet. Pharmacol. Therap. (in press).

69. Delatour, P., Garnier, F., Benoit, E., Longin, Ch., 1984. A correlation of toxicity of albendazole and oxfendazole with their free metabolites and bound residues. J. Vet. Pharmacol. Therap. (in press).

70. Craddock, V.M., Mattocks, A.R., Magee, P.N., 1968. The fate of 7(14-C)-methylguanine after administration to the rat. Biochem. J., 109, 75-78.

71. Jaggi, W., Lutz, W.K., Lüthy, J., Zweifel, U., Schlatter, C., 1980. In vivo covalent binding of aflatoxin metabolites isolated from animal tissue to rat-liver DNA. Food Cosmet. Toxicol., 18, 257-260.

72. King, C.R.G., Wearer, S.A., Narrod, S.A., 1965. Anthistamines and teratogenicity in the rat. J. Pharmacol. exp. Therap., 147, 391-398.

73. Posner, H.S., Graves, A., King, C.T.G., Wilk, A., 1967. Experimental alteration of the metabolism of chlorcyclizine and the indicence of cleft palate in rats. J. Pharmacol. exp. Therap., 155, 494-505.

74. Delatour, P., 1983. Evaluation of drug residues in animal tissues. In : Ruckebusch et al., Veterinary Pharmacology and Toxicology, MTP Press Limited, pp. 659-670.

75. Delatour, P., Boisseau, J., Burgat, V., Mergier, P., Milhaud, G., Richou-Bac, L., Rico, A., Siou, G., 1981. Schéma général d'évaluation de la sécurité des résidus de médicaments embryotoxiques. Ann. Rech. Vet. 12, 215-218.

76. Theodorides, V.J., DiCuollo, C.J., Nawalinski, T., Miller, C.R., Murphy, J.R., Freeman, J.F., Killeen, J.C., and Rapp, W.R., 1977. Toxicologic and teratologic studies of oxibendazole in ruminants and laboratory animals. Am. J. Vet. Res., 38, 809-814.

77. Speirs, A.L., 1962. Thalidomide and congenital abnormalities. Lancet. i.
        303-305.
78. James, L.F., Keeler, R.F., Binns, W., 1968. Teratogenic effects of aminop-
        terin in sheep. Teratology, 1, 407-412.

RESUME

EMBRYOTOXICITE ET TERATOGENESE
CONSEQUENCES POUR LES MEDICAMENTS VETERINAIRES

Paul DELATOUR,

Laboratoire de Biochimie,

Ecole Nationale Vétérinaire de Lyon,

F-69752 Charbonnières Cedex, France

Les manifestations tératologiques font partie des effets secondaires de nombreux médicaments et pesticides. Cette forme particulière de toxicité implique deux types de conséquences : l'existence possible d'une marge de sécurité pour l'espèce animale indiquée au cours de la période de gestation et un risque possible lié à la présence de résidus de ces produits dans les denrées d'origine animale (lait, viande ou oeufs) destinées à la consommation humaine.

L'étude des lois générales de la tératogénèse, valables tant chez l'homme que chez les animaux, met en évidence quelques aspects particuliers, qui sont fort importants pour nous : la plus ou moins grande sensibilité des espèces animales, la légèreté relative du stimulus tératogène, l'effet d'une dose unique, la notion de calendrier tératologique, la notion de dose sans effet, le rôle du métabolisme.

D'autre part, le raffinement des méthodes analytiques, qui rend possible la séparation des "résidus totaux" en différentes fractions résiduelles, ainsi que les données sur la relation entre le métabolisme et la tératogénèse indiquent que le risque tératologique n'est pas imputable généralement à toutes les fractions résiduelles. Dans chaque cas, la distinction entre les métabolites toxiques à risque réel et les résidus apparemment non incriminés dans le processus tératogène doit être faite. Ceci permet de définir des doses sans effet, des doses journalières acceptables et des délais d'attente afin d'assurer la protection de l'utilisateur et du consommateur.

Le risque tératologique de résidus de médicaments vétérinaires chez la femme enceinte ne peut être négligé ni exagéré : après dix années d'expérience dans le domaine de la tératologie et du métabolisme, nous pouvons raisonnable-ment supposer qu'il ne s'agit pas d'un risque majeur.

Des discussions entre spécialistes de plusieurs disciplines seront toujours utiles et bénéfiques, en particulier dans un domaine aussi sensible que celui-ci. Cependant, les biologistes et analystes revendiquent la respon-sabilité de l'évaluation toxicologique des résidus. Il est toujours possible d'exiger des marges de sécurité plus importantes et de plus longs délais d'at-tente. Il faut également se rappeler qu'une attitude raisonnable est fondée sur la mise en balance des risques et des bénéfices. Demander une nourriture vraiment naturelle équivaudrait à un retour au Moyen-Age. L'homme de Neanderthal n'a jamais été contaminé par des résidus de produits chimiques, mais vraisemblablement par des mycotoxines : il est mort à l'âge de 25 ans !

ZUSAMMENFASSUNG

EMBRYOTOXIZITÄT UND TERATOGENESE

FOLGERUNGEN FÜR TIERÄRZTLICHE HEILMITTEL

Paul DELATOUR,

Laboratorium für Biochemie,

Ecole Nationale Vétérinaire de Lyon,

F-69572 Charbonnières Cedex, Frankreich

Teratologische Erscheinungen gehören zu den Nebenwirkungen zahlreicher
Heilmittel und Schädlingsbekämpfungsmittel. Diese besondere Form von Toxizität schließt zwei Arten von Folgen in sich ein : das mögliche Vorhandensein
einer Sicherheitsmarge bei der betreffenden Tierart während der Tragzeit und
ein mögliches Risiko im Zusammenhang mit der Gegenwart von Rückständen dieser
Erzeugnisse in Nahrungsmitteln tierischen Ursprungs (Milch, Fleisch oder Eier),
die für den menschlichen Verzehr bestimmt sind.

Die Untersuchung der sowohl beim Menschen als auch bei Tieren gelten-
den allgemeinen Gesetze der Teratogenese zeigt einige besonderen Aspekte auf,
die für uns sehr wichtig sind : die größere oder weniger große Empfindlichkeit
der Tierarten, die relative Leichtigkeit des teratogenen Stimulus, die Wirkung
einer Einzeldosis, der Begriff des teratologischen Kalenders, der Begriff der
wirkungslosen Dosis, die Rolle des Metabolismus.

Außerdem lassen die Verfeinerung der analytischen Methoden, die eine
Aufspaltung der "Gesamtrückstände" in verschiedene Rückstandsfraktionen ermöglicht, sowie die Angaben über die Beziehung zwischen dem Metabolismus und der
Teratogenese darauf schließen, daß das teratologische Risiko nicht generell
allen Rückstandsfraktionen zuzuschreiben ist. In jedem Fall ist die Unter-
scheidung zwischen den toxischen Metaboliten mit echtem Risiko und den in
teratogenen Prozeß mutmaßlich nicht inkriminierten Rückständen vorzunehmen.
Damit lassen sich wirkungslose Dosierungen, annehmbare Tagesdosierungen und
Wartezeiten zum Gewährleisten des Schutzes der Benutzer und der Verbraucher
festlegen.

Das teratologische Risiko der Rückstände aus tierärztlichen Heilmitteln
bei schwangeren Frauen darf weder außer Acht gelassen noch übertrieben werden :
nach zehnjähriger Erfahrung im Bereich der Teratologie und des Metabolismus
dürfen wir begründeterweise annehmen, daß es sich nicht um ein größeres Risiko
handelt.

Debatten zwischen Fachleute verschiedener Fachrichtungen sind jedoch
immer nützlich und hilfreich, besonders in einem so heiklen Bereich. Die
Verantwortung für die toxikologische Beurteilung der Rückstände wird jedoch
von den Biologen und Analytikern beansprucht. Es besteht immer die Möglichkeit, umfangreichere Sicherheitsmargen und längere Wartezeiten zu fordern.

Man muß auch bedenken, daß eine vernünftige Einstellung auf dem Abwägen der Risiken und der Vorteile beruht. Die Forderung einer wirklich natürlichen Nahrung würde einen Rückfall ins Mittelalter bedeuten. Der Neandertaler wurde zwar nie von Rückständen chemischer Produkte kontaminiert, dafür aber wahrscheinlich von Pilzgiften : er starb im Alter von 25 Jahren !

ANABOLIC AGENTS : EVALUATION OF

HORMONO-MIMETIC AGENTS FOR MUTAGENIC

AND CARCINOGENIC POTENTIAL

Francis J.C. ROE, D.M., D.Sc., F.R.C. Path.

Introduction

The use of anabolic agents in meat production is mainly of interest to the toxicologist because of the possibility that residues of the agents used may remain in meat and meat products consumed by humans. The justification of using anabolic agents in meat production relates to the efficiency of meat production and the price of meat to the consumer. Residues from the use of anabolics in meat offer no direct health benefit to the consumer and therefore have the same status as contaminants.

The approach of Regulatory Authorities to the presence of contaminants in food is that the level should be at least 100 times less than a no-toxic-effect level in a sensitive species and that it should be as low as can be reasonably achieved by good agricultural practice. However, if a contaminant were found to be a genotoxic carcinogen, then Regulators would move to prevent any level of contamination of food with it.

Interference with farm animals by castration and/or the administration of anabolic agents is an integral part of present day farming and meat produc-tion, and it is clear that these practices are entirely safe from the view-point of the animals themselves. The question that I shall address concerns the safety of these practices for the consumer of meat derived from animals treated with anabolic agents.

During the last 5 - 10 years there has been intensive debate of this subject and in four areas a consensus view has been reached. Firstly, the exposure of animals or humans to high doses of either naturally-occurring oestrogens (e.g. 17β-oestradiol) or naturally-occurring androgens (e.g. testosterone) may increase the risk of development of certain specific forms of cancer. Secondly, these effects only occur under conditions of manifest disturbance of hormonal status and such agents are not genotoxic carcinogens. Thirdly, when such agents are properly used in meat production, the active residues of them that remain in meat are negligible in comparison with the levels of the same endogenously produced hormones that are present in the blood and tissues of normal human beings. This is true even for young children, prepubertal boys and post menopausal women. For this reason, there is now no concern about the proper use of natural steroids in meat production. Finally, it is now generally agreed that the oestrogenic stilbenes, such as diethylstilboestrol, should not be used in meat production. The reason is not because there is any proven risk that they can produce cancer other than by a non-genotoxic, hormonal mechanism, but because these agents are not readily deactivated by the liver. This poor capacity of the liver to deactive stilbenes means that residues of these agents in meat are much more likely to reach hormone-receptor sites in the tissues of meat-eaters than are residues of the steroids which are readily deactivated in the liver.

Remaining for debate is the safety of using non-stilbene xenobiotic anabolic agents. The two most important of these are trenbolone acetate (TBA), an androgenic steroidal agent with close structural similarity to testosterone and zeranol, a non-steroidal agent with oestrogenic properties.

As there are rather more data for TBA than for zeranol, I will devote most of the rest of my talk to this one substance.

Trenbolone acetate (TBA)

In the course of my brief review of trenbolone acetate (TBA), I will endeavour to answer four questions :

1.   Is there adequate evidence that neither TBA itself nor any residues derivable from it are genotoxic ?

2.   Is there adequate evidence from long-term in vivo studies in animals that TBA and its metabolites do not produce tumors other than as a consequence of their hormonal activity under conditions of excessive dosage ?

3.   Has a no-hormonal effect level been adequately demonstrated in a suitably sensitive species ?

And if the answer to the second question is positive

4.   Is there an adequate safety margin between the levels of residues of TBA and its metabolites in meat and the established no-hormonal-effect level ?

1.   Is there adequate evidence that neither TBA itself or any residues derivable from it are genotoxic ?

A priori one would not expect a steroid that is structurally and biologically similar to testosterone either to possess electrophilic properties itself or to be metabolised to an electrophile. Thus, it would be surprising if TBA or its main metabolites, namely 17α -hydroxytrenbolone (17α -OH trenbolone) and 17β -hydroxytrenbolone (17β -OH trenbolone) had given positive results in the Ames test or in any in vitro or in vivo test for clastogeni-

city. On the other hand, as John Ashby has pointed out (Richold, 1983),
hormones which have been found, when given in excessive dosage, to increase
the incidence of tumours, have also been found to be capable of transforming
mammalian cells in vitro in the absence of metabolizing enzymes (e.g. S-9 mix)
and to give borderline results in the in vitro gene-mutation test based on the
use of L5178Y mouse lymphoma cells.

In the light of the data summarized in Table 1, it is reasonable to
conclude that TBA, and the main metabolites of TBA which occur as residues in
meat give similar results to natural steroids in various laboratory tests for
mutagenicity and clastogenicity and that the use of TBA in meat productions
poses no genotoxic threat to the meat consumer.

2.   Is there adequate evidence from long-term in vivo studies in animals
that TBA and its metabolites do not produce tumours other than as a consequence of hormonal activity under conditions of excessive dosage ?

The hormonal activity of TBA is similar to, but not identical with,
that of testosterone. Thus, castrated cattle treated with TBA are less
aggressive than castrated cattle treated with testosterone. Also, whereas
testosterone stimulates protein synthesis, TBA achieves the same end point by
slowing the breakdown of protein (Sinnett-Smith et al, 1983 , Buttery and
Sinnett-Smith, 1984). According to Donaldson et al (1981), treatment of
cattle and sheep with testosterone leads to increases in plasma levels of
growth hormones whereas TBA has no such effect. Despite these differences in
mode of action, it is reasonable to use testosterone as a model for assessing
the hormone-associated activity of TBA.

Long-term feeding studies on TBA have been carried out in rats and
mice. The rat study (Hunter et al, 1982) included an in utero phase and
involved groups of 50-52 male and female Sprague-Dawley rats that were exposed
to 0, 0.5, 1, 4, 16 or 50 ppm TBA admixed with a standard laboratory chow.

The principal effects are summarized in Table 2 and details of the
effects on tumour incidence are given in Tables 3 and 4. The lowest level of
TBA tested, namely, 0.5 ppm in the diet, proved to be one at which no effects
of any kind were seen in rats of either sex. In males, the only significant
effects of treatment on tumour incidence, after correction for survival
differences between the groups were reductions of C-cell tumours of the
thyroid (50 ppm group) and of adrenal medullary tumours (16 ppm group). Both
of these differences may well have been due to chance. In females, highly
significant (p< 0.001) reductions in mammary and pituitary tumours were seen
in the 16 and 50 ppm groups. These and certain other effects are only what
one would expect from high level exposure of females to male sex hormones such

TABLE 1

Comparison of TBA and its metabolites with testosterone and 17β-oestradiol in various tests for mutagenicity and clastogenicity.

Test	TBA	17α-OH Trenbolone	17β-OH Trenbolone	Testosterone	17β-oestradiol	References
Salmonella (Ames)	–	–	–	NT	NT	Hossack et al, 1978. Ingerowski et al, 1981 Lang and Redmann, 1979.
Mammalian cell gene mutation (L5178Y mouse lymphoma)	NT	±	±	±	–	Drevon et al, 1981 Richold et al, 1982.
In vitro cytogenetics (human lymphocytes Chinese hamster cells)	NT	–	–	NT	NT	Ishidata and Odashima, 1977 Richold et al, 1982.
In vivo cytogenetics (rat testis and bone marrow)	–	–	–	–	NT	Richold and Richardson, 1982.
In vivo DNA covalent Binding index (CBI)*	5.61**	NT	NT	4.8**	11.4**	Barraud et al, 1983.

* CBI = $\dfrac{\mu\text{mole/mole DNA}}{\text{mmole of administered substance/kg body weight}}$

** Dose of TBA given = 17 ug/kg; testosterone = 19 ug/kg; 17 -oestradiol = 15 ug/kg

TABLE 2

Effects of life time dietary exposure to TBA (starting _in utero_) on Sprague-Dawley rats (from Hunter et al, 1982)

ppm in diet	Effect
4-50	Increased growth in $\female$
1-50	Impaired reproductive performance
1-50	Hormonal effects on $\female$ genitalia with secondary urinary obstruction and bladder changes at 16 and 50 ppm.
50	Non-significant excess of pancreatic islet-cell tumours
50	Significant reduction in mammary tumours and non significant reduction in pituitary tumours in $\female$
0.5	No-toxic-effect level and no-hormonal-effect level in both sexes.

TABLE 3

Effects of TBA on survival and age-standardized tumour incidence in male Sprague-Dawley rats (from Hunter et al, 1982, with analysis by Lee, 1983).

Level of TBA in diet (ppm)	0	0.5	1	4	16	50
% survival to 2 years	26	24	28	20	24	40^{+}
Thyroid - C-cell		NS	NS	NS	NS	-
Adrenal - Medulla		NS	NS	NS	-	NS
Any site		NS	NS	NS	NS	NS

NS = not significant
+ = more than controls, $p < 0.05$
- = less than controls, $p < 0.05$

TABLE 4

Effects of TBA on survival and age-standardized tumour incidence in female Sprague-Dawley rats (from Hunter et al, 1982, with analysis by Lee, 1983).

Level of TBA in diet (ppm)	0	0.5	1	4	16	50
% survival to 2 years	30	20	22	32	68	52^{+++}
Pancreas islet-cell		NS	NS	+	NS	+
Mammary		NS	NS	NS	---	---
Pituitary		+	NS	NS	NS	-
Any site		NS	NS	NS	---	---
Any site other than pituitary or mammary		NS	NS	NS	NS	NS

NS = not significant
+ = more than controls, $p < 0.05$ +++ = more than controls, $p < 0.001$
- = less than controls, $p < 0.05$ --- = less than controls, $p < 0.001$

as testosterone or TBA. It is possible that the equivocal effect of TBA on the incidence of islet-cell tumours of the pancreas was real since several investigators have reported associations between testosterone administration and islet-cell neoplasia (Neelon et al, 1975) and between growth hormone administration and islet-cell hyperplasia (Cavallero and Dora, 1954; Hausberger, 1961, Martin et al, 1968, Gepts et al, 1960, Bates and Garrison, 1974). In the United States the Cancer Assessment Committee reviewed the data for pancreatic islet-cell tumours and concluded "... the increased incidence of pancreatic islet-cell tumours in the Sprague Dawley female rat is not the result of a carcinogenic effect of trenbolone acetate" (CAC, 1983).

For the long-term study in mice (Hunter et al, 1981), groups of 5 male and 5 female Swiss mice were exposed to 0, 0.5, 1, 10 or 100 ppm TBA in diet for up to 104 weeks in the case of male and 95 weeks in the case of females. In response to the top two dose levels, there was, in males but not in females, a significantly increased incidence of nodular hyperplasia and/or liver-cell neoplasia. At all dose levels, TBA significantly enhanced body growth, but there was no adverse effect on life or health in animals exposed to 1 ppm TBA or less. At the lowest dose level tested, there was, in fact, a significant reduction in incidence of malignant tumours for all sites combined. The results are summarized in Tables 5 and 6.

There are three reasons why the increased incidence of liver tumour in response to TBA is not a matter for real toxicological concern. Firstly, the incidences seen in the study were within the range for historical untreated control mice of the same strain at the Huntingdon Research Centre. Secondly, there is abundant published evidence that androgen status influences liver tumour risk in mice. Such tumours are more common in males than females and their incidence can be decreased in males by castration and increased by the administration of androgens (Agnew and Gardner, 1952, Andervont, 1951, Taylor, 1983). Thirdly, it is clear that high food consumption and consequential high growth rate predispose non-specifically to increased liver tumour incidence in mice (Tucker, 1979, Conybeare, 1980).

One may safely conclude that, if the effect on the incidence of liver tumours in this study was real, then it was a non-specific response to androgen treatment and/or growth promotion, and not a manifestation of genotoxic carcinogenicity. This was the conclusion of the Cancer Assessment Committee in the USA who reported : "... the increase in liver tumours in male mice is attributable to the hormonal effect of this agent" (CAC, 1983).

TABLE 5

Effects of life-time exposure of Swiss mice to TBA in the diet (from Hunter et al, 1981).

ppm in diet	Effects
10-100	Reduced survival in males, improved survival in females.
0.5-100	Increased growth – only marginal at 0.5 and 1 ppm.
100	Inhibition of ovulation and effects on uterus and preputial glands.
10-100	Increased incidence of nodular hyperplasia and of neoplasia in liver in males.

TABLE 6

Effects of TBA on survival and tumour incidence in Swiss mice (from Hunter et al, 1981)

Males

ppm TBA in diet	0	0.5	1	10	100
% survival to 96 weeks	44	35	42	37	25
% liver tumours	17	27	27	33^+	37^{++}
% malignant tumour at any site	38	33	38	44	34

Females

% survival to 96 weeks	21	25	27	35	38
% liver tumours	4	2	6	10	16^{++}
% malignant tumour at any site	65	44^-	44^-	47	46

$+$ = more than controls, $p < 0.05$

$++$ = more than controls, $p < 0.01$

$+++$ = more than controls, $p < 0.001$

$-$ = less than controls, $p < 0.05$

132

3.   <u>Has a no-hormonal-effect level for TBA been adequately demonstrated in a suitably sensitive species ?</u>

As I have indicated above, 0.5 ppm in the diet was a completely no-effect level in the 2-year rat study conducted at the Huntingdon Research Centre. In the long-term mouse study, no adverse effects occurred at this level or at 1 ppm, but there was evidence of slight growth promotion in females even at the 0.5 ppm dietary level. Arguably, growth promotion represents a beneficial hormonal effect. In any case, I believe it is safe to assume that 0.1 ppm in the diet is a completely no-hormonal-effect level for the mice.

In the male pig the epithelium of the prostate is extremely sensitive to changes in androgen status. Castration leads to acinar atrophy and to the epithelial cells becoming low cuboid in shape. At the same time, the fibro-muscular stroma becomes more prominent. Administration of testosterone corrects these changes in proportion to dose. In castrated pigs, the admi-nistration of 17α -OH TBA in doses ranging from 0.1 to 360 μg/kg/day had no discernible effect on the epithelium of the prostate. Similarly, the daily administration of 0.1 to 10 μg/kg/day 17β -OH TBA was without effect. Only when the daily dose was increased to 16, 24 or 36 μg/kg/day were epithelial changes evident (Roberts et al, 1983).

In the same animals, the rise in the level of circulating plasma luteinizing hormone (LH) which normally occurs during the 14 days after castration was delayed in animals receiving 16, 24 or 36 μg/kg/day 17β -OH TBA or any dose of 17α -OH TBA up to 360 μg/kg/day.

These observations are consistent with 10 μg/kg/day being a no-hormonal for 17β -OH TBA and 360 μg/kg/day being a no-hormonal effect for 17α -OH TBA in the pig.

In a study on female rhesus macaque monkeys (Hess et al, 1984) three groups of 6 animals were fed on diets providing 60, 240 or 960 μg TBA/day. Blood levels of various circulating hormones in samples taken regularly during 3 menstrual cycles after the start of exposure were compared with levels during a pretreatment menstrual cycle. The hormones assayed were oestradiol, progesterone, luteinizing hormone, and follicle-stimulating hormone (FSH). Equivocal effects on FSH and ovarian function were seen in animal exposed to the highest dietary concentration of TBA, but the middle and low dose levels (equivalent to 40 μg/kg/day and 10 μg/kg/day, respectively) were without discernible effect on these hormones.

4.   Is there an adequate safety margin between the levels of residues of
TBA and its metabolites in meat and the established no-hormonal-effect level ?

In the light of the available data on residue levels and on
no-hormonal-effect levels in various species, the 1983 Joint FAO/WHO Expert
Committee on Food Additives (JEFCA) concluded that the safety margin is
adequate. The report (WHO, 1983) reads : "The Committee was of the opinion
that the low residue levels of trenbolone acetate and its metabolites in meat
products would result in exposures far below the levels at which hormonal
activity was observed in animal models." In the light of this opinion, the
Committee provisionally accepted the use of trenbolone acetate as an anabolic
agent for the production of meat for human consumption in accordance with good
laboratory practice. Subsequent to this decision, the results of the study by
Hess (1984) in primates (vide supra) have become available. These should, in
due course, be adequate to remove the 'provisional' status of the JEFCA
opinion.

Zeranol

A series of tests of zeranol for mutagenicity and clastogenicity are
summarized in Table 7. The negative results in these tests provide adequate
assurance that zeranol does not damage DNA.

At one time, zeranol was subjected to various short and long term
tests in animals because it was thought that it might be of value as an oral
contraceptive. Thus, there are available for scrutiny the results of a 90-day
toxicity study in rats, a 2-year chronic toxicity/carcinogenicity study in
rats, a 7-year study in female dogs, a 10-year study in female rhesus monkeys
and a 3-generation reproduction/teratology study in rats. Between them these
studies provide considerable reassurance of the safety of zeranol since the
only effects seen were attributable to its known oestrogenicity. More recently, 0.025 mg/kg/day has been found to be a no-toxic-effect level in a 90-day
rat study and, using vaginal cornification as the end-point (N.B. arguably the
most sensitive detector of oestrogenic activity), 0.05 mg/kg/day has been
found to be a no-effect level for zeranol in the monkey. However, some
regulatory authorities who have considered the data for zeranol have required
the results of carcinogenicity studies in rats and mice conducted according to
present-day standards. The manufacturers are presently sponsoring studies to
fulfil these requirements.

On the basis of presently available data, there are no grounds for
anticipating any toxic hazard to meat consumers from residues in meat
originating from zeranol-treated animals.

TABLE 7

Results of mutagenicity and clastogenicity tests on zeranol.

Test	Details	Result	Reference
Salmonella (Ames)	5 tester strains $\pm$ activation	Negative	Litton Bionetics Report No. 20988 (October 1982)
Mammalian cell gene mutation (L5178Y mouse lymphoma)	up to 600 μg/ml without activation	Negative	Litton Bionetics Report No. 20989 (December 1982)
	up to 100 μg/ml without activation	Negative	
In vivo clastogenicity (mouse bone marrow)	up to 5 g/kg/day	Negative	Litton Bionetics Report No. 22202 (December 1982)
Hepatocyte Primary Culture/DNA repair assay	up to 0.013 mg/ml	Negative	G.M. Williams, Report to IMC (12 December 1983)
In vitro covalent binding to DNA	1.65 for zeranol compared with 11.4 for 17β-oestradiol		Barraud et al (1983)

Conclusions

As far as TBA is concerned, the answers to the four crucial questions listed at the start of this paper are as follows.

1.   There is adequate evidence that TBA and its metabolites are not genotoxic.

2.   There is no evidence that TBA causes tumours in animals other than by a hormonal mechanism when given in excessive dosage.

3.   0.1 – 0.5 ppm in the diet has been adequately shown to be a no-hormonal-effect level in mice and rats. 10 µg/kg/day in the diet has been found to be no-hormonal-effect level in the mature female pig and 40 µg/kg/day has been found to be so in the female rhesus monkey.

4.   There is a large and adequate safety margin between levels of residues known to be without hormonal effect in various species of animal (rat, mouse, pig and rhesus monkey) and the levels of exposure of humans to residues in meat from treated animals.

Summary

1.   Residues of xenobiotic anabolic agents or their metabolites in meat are to be regarded as contaminants and in the same category as residues of veterinary medicines or pesticides.

2.   In the case of non-genotoxic xenobiotic substances, residues may be accepted as posing no toxic or carcinogenic threat to man, provided that there is an adequate safety margin between the level in food and an established no-toxic effect level in a suitably sensitive species. Nevertheless, the maximum level of a residue permitted should allow for the highest safety margin consistent with good agricultural practice.

3.   Trenbolone acetate (TBA) is a synthetic steroid with androgenic activity. There is adequate evidence that neither TBA itself nor its main metabolites are genotoxic. In the opinion of the author and of Cancer Assessment Committee in the USA, such effects as high doses of TBA have on tumour incidence in rats and mice are secondary to its hormonal activity. There is no increase in tumour risk where exposure is below the level for the manifestation of hormonal activity. The Joint FAO/WHO Expert Committee on Food Additives have accepted that evidence from studies on rats, mice and pigs establish no hormonal effect levels in rats, mice and pigs and that there is an adequate safety margin between these levels and the levels of residues of TBA and its metabolites in meat from treated animals. After this view was reached further data indicating a no hormonal effect level of 40 µg/kg/day in the diet in the rhesus monkey have become available.

136

4.   Existing data establish that zeranol is not genotoxic or teratogenic.
Carcinogenicity studies in rats and mice are in progress. On the basis of
presently available data there are no grounds for suspecting any toxic hazard
to meat consumers from residues in meat originating from zeranol-treated
animals.

References

Agnew, L.R.C., and Gardner, W.U., 1952. The incidence of spontaneous hepatomas
    in C_3H, C_3H (Low Milk Factor), and CBA mice and the effect of estrogen and
    androgen on the occurrence of these tumours in C_3H mice. Cancer Res., 12 :
    757-761.
Andervont, 1952. Studies on the occurrence of spontaneous hepatomas in mice
    of strains C_3H and CBA. J. Natl. Cancer Inst., 11 : 581-592.
Barraud, B., Lugnier, A., and Dirheimer, G., 1983). In vivo covalent binding
    to rat liver DNA of trenbolone as compared with 17 -oestradiol, testos-
    terone and zeranol. Chapter in : Anabolics in Animal Production, published
    by O.I.E., pp. 193-206.
Bates, R.W., and Garrison, M.M., 1974. Daily changes in concentration of pan-
    creatic and serum insulin and of blood glucose during 5 days of treatment
    of rats with growth hormone, ACTH, Cortisol, dexamethasone and tolbutamide
    alone and in combinations. Metabolism, 23 : 947.
Buttery, P.J. and Sinnett-Smith, P.A., 1984. The mode of action of anabolic
    agents. In : Manipulation of growth in farm animals, edited by J.F. Roche,
    D. O'Callaghan and D. Martinus-Nijhoff.
Cancer Assessment Committee, 1983. Memorandum from Dr. T.M. Farber to
    Dr. M. Norcross, Associate Director of Food Safety U.S. Dept. of Health
    and Human Services, 29 August, 1983.
Cavallero, C., and Dova, E., 1954. Morphological changes in islets of
    langerhans of the pituitary of dwarf mice during induced growth. Acta
    Path. Microbiol. Scand., 34 : 201.
Conybeare, G., 1980. Effect of quality and quantity of diet on survival and
    tumour incidence in outbred Swiss mice. Fd. Cosmet. Toxicol., 18 : 65-75.
Donaldson, I.A., Hart, I.C. and Heitzman, R.J., 1981. Growth hormone,
    insulin, prolactin and total thyroxin in the plasma of sheep implanted
    with the anabolic steroid, trenbolone acetate, alone or with oestradiol.
    Res. Vet. Sci., 30 : 7.
Drevon, C., Piccoli, C., and Montesano, R., 1981. Mutagenicity assay of
    oestrogenic hormones in mammalian cells. Mutation Res., 89, 83-90.
Gepts, W., Christophe, J. and Mayer, J., 1960. Pancreatic islets in mice
    with the obese-hyperglycemic syndrome. Lack of effect of carbutamide.
    Diabetes, 9 : 63.
Hausberger, F.X., 1961. Effects of food restriction on body composition and
    islet hypertrophy of mice bearing corticotrophin-secreting tumours.
    Acta Endocrinol., 37 : 336.
Hess, D.L., 1984. Determination of the hormonal no-effect dose level for
    trenbolone acetate in the female rhesus macaque. Report from the Oregon
    Primate Centre, dated 9 March 1984.
Hossack, D.J.N., Richold, M., Jones, E., 1978. Ames metabolic activation
    test to assess the potential mutagenic effect of trenbolone acetate,
    17 -oestradiol and a 7:1 mixture of trenbolone with 17 -oestradiol.
    Huntingdon Research Centre Report No. RSL/351/781136.
Hunter, B., Graham, C., Gibson, W.A., Read, R., Gregson, R., Heywood, R.,
    Offer, J., Prentice, D. and Woodhouse, R., 1981. Long-term feeding of
    trenbolone acetate to mice. Huntingdon Research Centre Report No. RSL/
    267/80122.
Hunter, B., Batham, P., Heywood, R., Street, A.E., Gibson, W., Prentice, D.,
    Gregson, R., Offer, J., and Almond, R.H., 1982. Trenbolone acetate

potential tumorigenic and toxic effects in prolonged administration to rats following 'in utero' exposure. Huntingdon Research Centre Report No. RSL 295-G/80974.

Ingerowski, G.H., Scheutwinkel-Reich, M., and Stan, H-J., 1981. Mutagenicity studies on veterinary anabolic drugs with the Salmonella/microsome test. Mutation Res., 91 : 93-98.

Ishidata, M., and Odashima, S., 1977. Chromosome tests with 134 compounds on Chinese hamster cells in vitro - a screening for chemical carcinogens. Mutation Res., 48 : 337-354.

Lang, R., and Redmann, U., 1979. Non-mutagenicity of some sex hormones in the Ames Salmonella/Microsome mutagenicity test. Mutation Res., 67 : 361-365.

Lee, P.N., 1983. Trenbolone acetate : statistical analysis of tumour data from long-term study in rats carried out at H.R.C.

Martin, J.M., Akerblom, K.H., and Garay, G., 1968. Insulin secretion in rats with elevated levels of circulating growth hormone due to MtT-W15 tumour. J. Amer. Diabet. Ass., 17 : 661.

Neelon, F.A., Delcher, H.K., Steinman, H.M. and Lebovitz, H.E., 1975. A comparison of the structure of hamster pancreatin insulin and the insulin extracted from a transplantable hamster islet-cell carcinoma. Biochim. Biophys. Acta., 412 : 1.

Richold, M., 1983. An evaluation of the mutagenicity of anabolic hormones with particular reference to trenbolone. In : Anabolics in Animal Production, O.I.E., pp. 179-188.

Richold, M. and Richardson, J.C., 1982. Metaphase analysis on 17- -hydroxy-trenbolone and 17- -hydroxytrenbolone. Huntingdon Research Centre Report No. RSL 512/81962.

Richold, M., Dehnel, J.M. and Ransome, S., 1982. An assessment of the mutagenic potential of four compounds using an in vitro mammalian cell test system. Huntingdon Research Centre Report No. RSL 513/8286.

Richold, M., Richardson, J.C., Proudlock, R.J. and Fleming, P.M., 1982. Effect of 17- -hydroxytrenbolone and 17- -hydroxytrenbolone on cultured human lymphocytes. Huntingdon Research Centre Report No. RSL 544/82136.

Roberts, N.L., Cameron, D.M., and Foxcroft, G.R., 1983. Effects of trenbolone on plasma LH levels and histological changes following castration of mature male pigs. Huntingdon Research Centre Report No. RSL/581.

Roe, F.J.C., 1981. Toxicological aspects of anabolic hormonal use in meat production. In : Anabolizzanti in Zootecnia e Salute pubblica', Proc. of a Congress organised by the Società Italiana di Buiatria, Rome, pp. 159-169.

Sinnett-Smith, P.A., Dumelow, N.W. and Buttery, P.J., 1983. The effects of trenbolone acetate and zeranol on protein metabolism in male castrate and female lambs. Brit. J.Nutrition, (in press).

Tucker, M.J., 1979. The effects of long-term food restriction on tumours in rodents. Int. J. Cancer, 23 : 803-807.

World Health Organisation, 1983. Evaluation of certain food additives and contaminants. 27th Report of the Joint FAO/WHO Expert Committee on Food Additives. Technical Report Series No. 696.

RESUME

AGENTS ANABOLISANTS : EVALUATION DU POUVOIR CARCINOGENE ET MUTAGENE DES
SUBSTANCES A ACTIONS HORMONO-MIMETIQUES

Francis J. C. Roe, DM, DSc, FRC Path
Member of the Institute of Cancer Research, London,
Independent Consultant in Toxicology

De nos jours, les carcinogènes sont communément classés dans deux
catégories : les carcinogènes génotoxiques et les non-génotoxiques (épigéné-
tiques). Ceux du premier groupe ou leurs métabolites endommagent le matériel
génétique des cellules (ADN) et ceci augmente le risque de cancer. Par con-
tre, les carcinogènes non-génotoxiques n'attaquent pas l'ADN ni les chromosomes, mais agissent de diverses manières pour favoriser le cancer, entre
autres par l'établissement d'un déséquilibre hormonal. De nombreuses hormones
naturelles sont connues pour, à des doses excessives, agir de cette façon et
augmenter le risque de l'une ou l'autre forme de cancer, bien que le déséquilibre, directement responsable d'un risque à l'un ou l'autre endroit du corps,
puisse, en même temps, favoriser une diminution d'un tel risque à un autre
endroit du corps. Aucune preuve n'a été apportée que les hormones naturelles
influencent le risque de cancer dans un sens ou dans l'autre à des niveaux
d'exposition inférieurs à ceux requis pour la production d'effets hormonaux.

Il faut s'attendre à ce que les principes synthétiques ayant un effet
hormono-mimétique aient les mêmes effets sur les risques de cancer que les
hormones naturelles qu'ils imitent. De même, il faut s'attendre à ce qu'ils
n'aient aucun effet sur ces risques à des doses inférieures à celles requises
pour obtenir un effet hormonal. Toutefois, des assurances supplémentaires
quant à la sécurité sont nécessaires pour les substances à effet hormono-mimétique, car elles sont différentes des hormones naturelles. Il est nécessaire
de s'assurer que ces substances et leurs métabolites ne soient pas des carcinogènes génotoxiques et qu'elles ne favorisent pas les cancers par des méca-
nismes non génotoxiques autres que ceux de leur activité hormono-mimétique.
Ces assurances supplémentaires sont absolument les mêmes que celles demandées
pour les produits chimiques ou autres, tels que résidus de pesticides, d'additifs pour l'alimentation et des moyens utilisés dans la fabrication d'aliments. Des tests sont disponibles pour obtenir ces assurances. Ce sont les
mêmes que ceux utilisés pour d'autres types de produits chimiques.

La position en ce qui concerne l'évaluation de la sécurité de deux
agents hormono-mimétiques utilisés en production animale, l'acétate de
trenbolone et le zeranol, est qu'ils feront l'objet de nouvelles études. De

même, le cas de l'utilisation du diethylstilbestrol (DES) et d'autres stil-
bènes sera discuté à nouveau. Quel que soit le statut de ces composés d'un
point de vue mutagénique ou carcinogénique, ils ne devraient pas être utilisés, simplement parce le foie les désactive mal, avec le résultat que leurs
résidus dans la viande peuvent atteindre des sites récepteurs dans le corps
des consommateurs.

ZUSAMMENFASSUNG

ANABOLE SUBSTANZEN : BEDEUTUNG HORMONWIRKSAMER SUBSTANZEN
ALS MUTAGENE UND CARCINOGENE GEFAHREN
Francis J.C. Roe, DM, DSc, FRC Path

Es ist heute üblich, Carcinogene in eine der beiden Kategorien einzuordnen : genotoxisch und nicht-genotoxisch (epigenetisch). Genotoxische Substanzen oder ihre Metabolite beschädigen das genetische Material der Zellen, die DNA. Ein solcher Schaden (Gen-Mutation oder chromosomale Abberration) kann, wenn er nicht rückgängig gemacht wird, zu einem erhöhten Krebsrisiko prädisponieren. Nicht-genotoxische Carcinogene beschädigen weder die DNA noch die Chromosomen, sondern erhöhen auf vielfältige andere Weise das Krebsrisiko. Einer dieser Wege hat zur Folge, daß sich hormonelles Ungleichgewicht einstellt. Viele natürliche Hormone sind dafür bekannt, daß sie in exzessiver Dosierung die Gefahr der einen oder anderen Krebsform erhöhen. Auf der anderen Seite kann dieses hormonelle Ungleichgewicht das Krebsrisiko an anderen Körperorganen reduzieren. Es hat sich gezeigt, daß kein natürliches Hormon, wenn es in geringeren Mengen vorhanden ist, als für die Auslösung der typischen Wirkungen vonnöten ist.

Synthetische Substanzen mit hormonähnlichen Wirkungen haben in übermässiger Dosierung die gleichen Auswirkungen auf die Krebsrisiken wie die natürlichen Hormone, deren Wirkungen sie imitieren. In ähnlicher Weise kann man erwarten, daß sie in einer Dosierung, die ihre typischen Reaktionen noch nicht auslöst, auch keine Auswirkungen auf die Krebsrisiken haben. Dennoch muß zusätzlich sichergestellt werden, daß solche Stoffe mit hormonähnlichen Wirkungen keine unerwünschten Nebeneffekte zeigen. Wir müssen sicher sein, daß sie und ihre Metabolite nicht zu den genotoxischen Carcinogenen zählen und daß sie auch nicht durch einen anderen nicht-genotoxischen Mechanismus, außerhalb ihrer hormonähnlichen Wirksamkeit, zum Krebs prädisponieren. Diese zusätzlichen Sicherheiten sind genau dieselben, die auch für andere Chemikalien gebraucht werden, z.B. für Pestizidrückstände, Nahrungsmittelzusätze und Stoffe, die bei der Verarbeitung von Nahrungsmittel eine Rolle spielen. In ähnlicher Weise sind die Tests zur Überprüfung dieselben wie für andere Gruppen chemischer Substanzen.

Die Stellung zweier Stoffe, die hormonähnliche Wirkungen ausüben und die heute allgemein bei der Fleischproduktion eingesetzt werden, nämlich des Trenbolonazetats und des Zeranols werden im Hinblick auf diese Sicherheitsbewertung behandelt. Ebenso wird die Untauglichkeit des Diethylstilböstrols (DES) und anderer Stilbene in der Tierproduktion diskutiert. Ganz unabhänglich davon ob sie nun mutagen oder carcinogen sind, sollten sie nicht verwendet werden, weil sie in der Leber nur wenig abgebaut werden. Das hat zur

Folge, daß Rückstände in Nahrungsmitteln Hormonrezeptoren im Körper des Konsumenten erreichen können.

<u>ALLERGY AND RESIDUES</u>

Vivianne BURGAT,

Laboratoire de Toxicologie Biochimique et Métabolique (INRA),

Ecole Nationale Vétérinaire de Toulouse,

F-31076 Toulouse Cedex, France

The residues of veterinary drugs or additives used in animal feeding-stuffs are sometimes incriminated in human allergies. To our knowledge, the observations made in this field only represent a small percentage of the food intolerance phenomena. Most reactions of this type, the aetiology of which is known, are indeed mainly attributed to normal food constituents.

However, the very sensitizing characteristic of some products and their accepted presence in food of animal origin may not exclude the effects of residues.

This problem will be discussed after a brief summary of the basic concepts involved. The last part of this presentation will discuss the evaluation and the control of the risk.

1.   <u>Immunological bases</u>

a) <u>Hypersensitivity : nature and symptoms</u>

The hypersensitivity is a form of immune response resulting from a specific antigenic pressure (fig. 1)

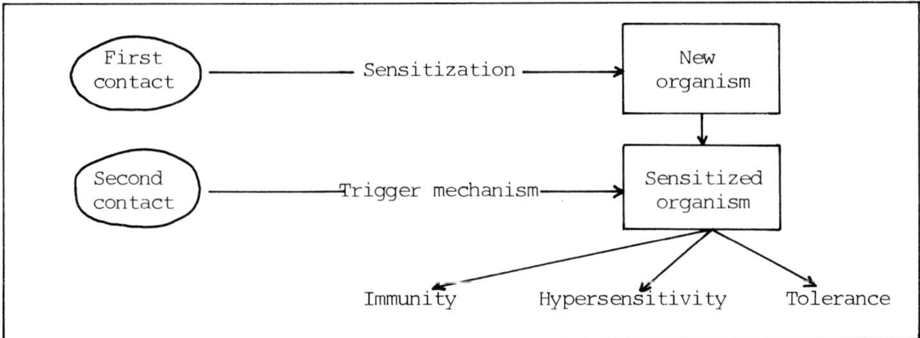

Fig. 1 : Position of hypersensitivity in the immune response.

144

Its symptoms are varied : four types of reaction are classically described (Fig. 2).

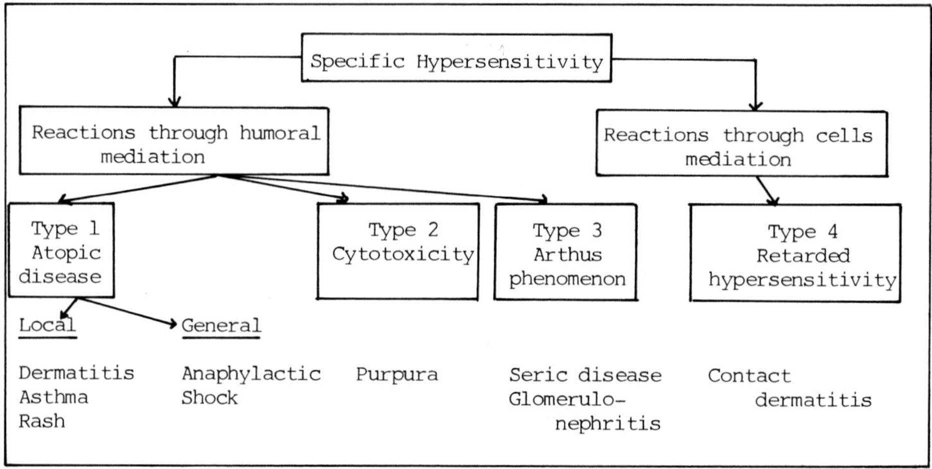

Fig. 2 : Hypersensitivity symptoms (Gell and Coombs).

Allergic accidents observed during therapeutic treatment in man can belong to type 1, 2, 3 or 4, varying according to the compound and the mode of contact. Observations concerning residues involved reactions of type 1 (immediate hypersensitivity or anaphylaxis).

b) <u>Immunogenicity - Allergenicity</u>
There is a profound uniformity in the mechanism of immune response. This response involves various elements, which are schematically represented in fig. 3 for the type 1 reaction.

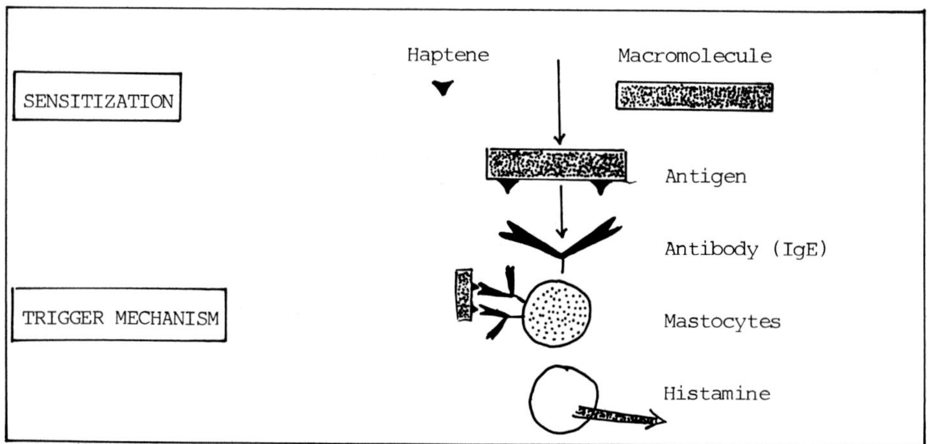

Fig. 3.

Medicinal products (haptenes) are not active in their native state because of their low molecular weight. Reactivity is only acquired after fixation on molecules with a high molecular weight (generally endogenous proteins) : the complex is called an antigen (Ag).

The immunogenicity of an Ag is its capacity of causing an immune response; in the present case of synthesis, the specific antibody (IgE).

The allergenicity of an Ag is linked to its ability of being recognized by antibodies (or some cell structures) which thus permit the triggering of a reaction of hypersensitivity in an individual who has first been sensitized.

Depending on their nature, Ags will preferentially have an effect on sensitization or the trigger mechanism. The case of Benzylpenicillin (BP) is taken as an example (fig. 4).

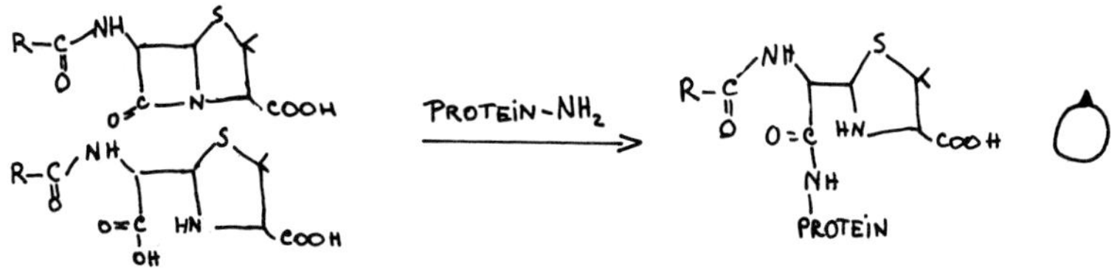

Antigen	Immunogenicity	Allergenicity
BP + autologous protein	+ 	+ 
BP + heterologous protein	++ 	
Polymer (BP)n	−	+
BP + polyol (glucose)	−	+

Fig. 4 : Formation of a penicillin antigen.

Penicillin antigens may be formed by :
- in vivo penicilloylation of autologous proteins. The reaction varies with the dose and the route of administration. So the number of BP molecules fixed on a protein (epitope) increases with the dose and is greater after intra-muscular than after intravenous administration. These antigens do not induce Ab synthesis until 30 epitopes (or more) are linked to an autologous protein;
- penicillin reaction with heterologous fungal proteins (preparation impurities 0.5 - 5 μg/g of penicillin). These penicilloyled proteins are more immunogenic than the abovementioned one : 6 epitopes per molecule are sufficient to provoke Ab synthesis;
- penicillin polymerisation (homopolymer of 5 to 15 units) or ester formation

146

between penicillin and polyols (glucose–dextrane); these derivates are highly
allergenic.

Conditions of sensitization and trigger mechanism are thus not completely similar. Sensitisation of the individual requires immunogenic antigens,
whereas the allergic reaction can be triggered by various conjugates
(Bundgaard, H., 1983).

2.   The problem of residues

a) Nature of the risk

The immunopathologic phenomena which are involved can a priori be of
two types : either the residues are sensitizing and can be the primary cause
of accidents observed later during other contacts (namely therapeutic) and the
risk is important; or the residues are triggering and can cause an hypersensitivity reaction in already sensitized individuals (fig. 5).

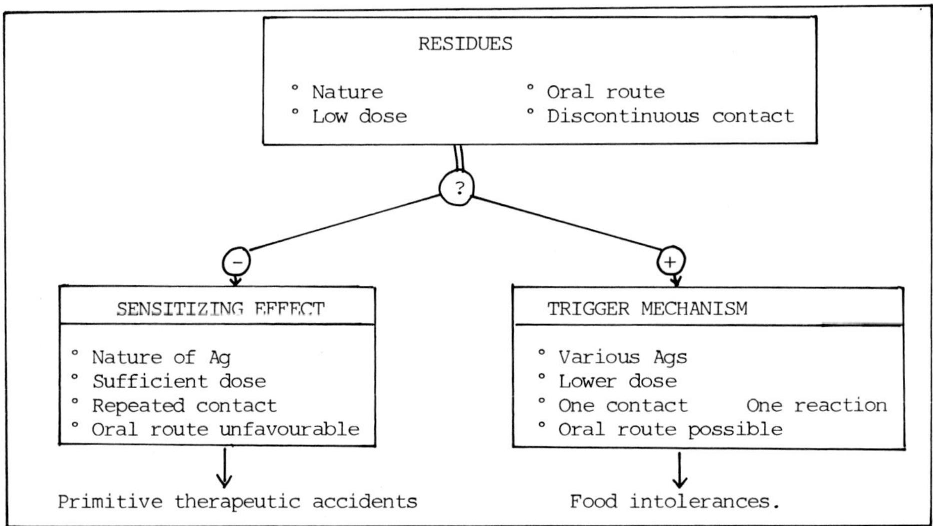

Fig. 5.

The importance of the two phenomena must be assessed in function of
the particular nature of the problem. These effects must be provoked by :
- low doses, since the residual contents are very low and are not related to
the dosages used in human medicine; they are in the order of the milligram per
kilo of food, and often lower. Regulation foresees for example that milk may
not contain antibiotics; in practice this means that if traces remain, they
will be below the detection limit, i.e. 0.005 I.U./ml (or 3 µg/l) for
penicillin;

- <u>products given by the oral route</u>, as these residues are found in human food. Now the immune response, whether in sensitizing or triggering phenomena, is generally more difficult to obtain by this route than by other modes of administration, as parenteral administration or inhalation. The active doses are notably higher, local defense mechanism may operate, and following repeated intake a tolerogenic response can even be obtained;

- a <u>discontinous contact</u>, as one cannot envisage that contamination be such that man eats daily the same type of residues. The problem here is different from that of additives authorized in human food. In this case, the intake is fortuitous and often unrepeated before several weeks, months, or years.

Various arguments plead a priori for a minimal effect of residues in the phenomenon of sensitization. On the contrary the trigger effect has been observed in human clinic.

b) <u>Sensitizing effect</u>

The proof of occult sensitization is not easy. One can either :
- detect sensitized individuals by appropriate test,
or
- study the allergic accidents observed in human medicine.

The first approach is not easy; it is at least extremely infrequent to our knowledge. We will for example quote the results of Ponvert et al. (1978) obtained with the test of leucocytotic migration (TLM) : in 131 persons recently untreated, 10 % of TLM are positive for tetracycline, 26 % for penicillin; on the contrary all tests were negative for sulfamethoxazol and glafenine. The authors suggest that there would be occult sensitization for penicillin and tetracycline (fig. 6).

	Positive TLM (%)	
	"Untreated"	"Treated"
Sulfamethoxazol	0	28
Glafenine	0	13
Tetracycline	10	16
Penicillin	26	45

Fig. 6.

Certain remarks are however necessary. The first concerns the definition of "recently untreated" individuals which does not exclude other earlier treatments. The second relates to occult sensitization; the residues of veterinary drugs may be incriminated, but there are also other causes which we will discuss later. Lastly, immunologists are far from being unanimous on the significance of this or other tests.

So cutanous tests with penicilloyl-polylysine are positive in 4 to 12 % of "non allergic" subjects, as well as specific hemagglutination tests in 30 % of patients treated with penicillin, even in the absence of any allergic reaction.

b) Another approach, an epidemiologic approach, can be envisaged. So we have tried to obtain information regarding the origin of the so-called primitive allergic accidents observed during the first therapeutic administration of a product. They result from occult sensitization and veterinary products may be incriminated.

Only one substance has actually be subjected to intensive studies, benzylpenicillin. Several enquiries at worldwide level show that there are approximately 0.7 to 1 % of allergic reactions after therapeutic administration of this particularly immunogenic substance. Among these cases of hypersensitivity 0.015 to 0.04 % are anaphylactic; 1 or 2 cases of lethal anaphylactic shocks occur for 100,000 treatments (fig. 7).

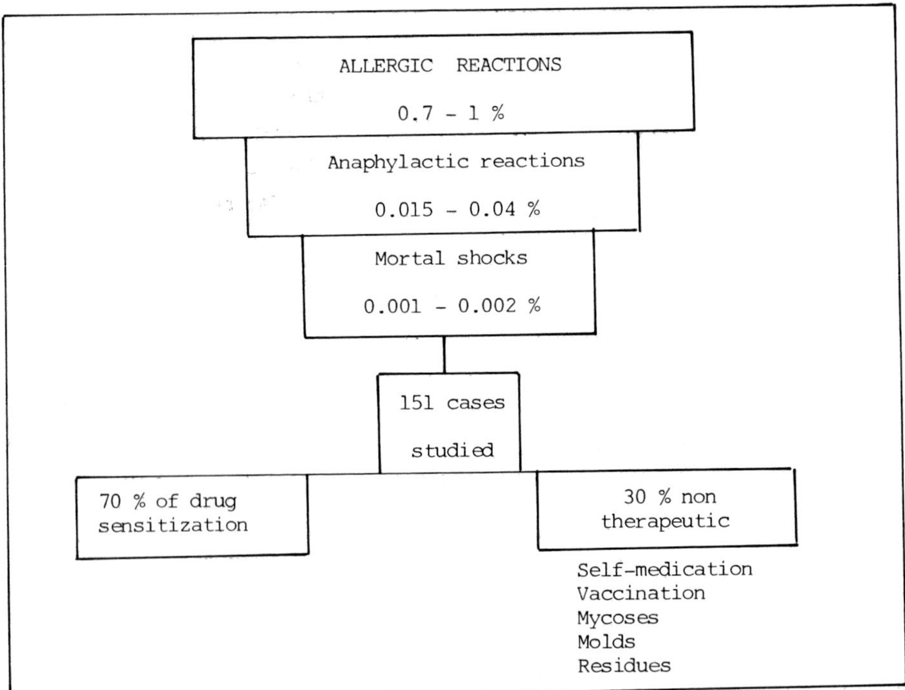

Fig. 7

Among the lethal shocks 151 cases have been particularly studied :
almost all have been caused by therapeutic doses administered intramuscularly;
there were 5 exceptions, two of these were observed after a trial test (0.01
I.U. via I.V., 4,000 I.U. intradermally), and three following oral treatment.
In addition 70 % of these shocks are secondary : the sensitization is attributable to an earlier antibiotic treatment given 10 days to 6-7 years prior to
the triggering treatment. In 30 % of mortal shocks, enquiry has not been able
to determine the origin of the sensitization, either because of lack of
information, or because these were occult sensitizations : automedication,
presence of penicillin (at that time) in the antipolyomyelitis vaccine, or
residues of penicillin (namely in milk); these causes are classically
mentioned in primitive accidents, but are far from being the only ones (Idsoe
et al, 1968).

So it is surprising that in the case of penicillin some occult
contaminations are hardly ever mentioned. Several molds or inferior fungi
secrete penicillin, such as Penicillium chrysogenum and Aspergilus flavus
which are currently present in salted meats, some cheese, powder milk, spoiled
fruit... A reaction of intolerance has been reported after intake of fruit
juice containing penicillin (Wicher, 1980).   A simple mycosis can
also have an effect : dermatophytes indeed secrete penicillin.

Where penicillin is concerned, severe allergic reactions are relatively rare and have been observed during therapeutic usage. Although one
cannot in the absolute exclude the role of residues of veterinary drugs, no
fact has to our knowledge formally or in any large manner involved them in
primitive sensitization.

In addition it must be said that the number of lethal anaphylactic
shocks due to penicillin has considerably decreased over the years; so, in the
U.K., 8 cases were recorded in 1957, 1 in 1964. This is attributed to the
improvement of the quality of the product used, as active immunogenic
substances are not only penicillin itself and its degradation product, but
also (and mostly) impurities.

c) Triggering effects

Reaction of food intolerance observed with food from animal origin to
which man is not allergic may show the triggering effect of a "non natural"
substance present in food.

These facts are well known in clinic. The residues of veterinary
drugs (essentially antibiotics) are sometimes incriminated, but in a
relatively low number of cases (a few percent) (Gounelle H. and Szakvary, A.,
1966).

The symptoms are not considered severe : rash, sometimes Quincke

150

oedema, but never, to our knowledge, an anaphylactic shock. It must actually
be noted that there is often confusion between "anaphylactic reaction"
(particularly used by anglicists) which is synonym of immediate
hypersensitivity, and "anaphylactic shock", which is a generalized reaction
that can be lethal.

In any case, these observations exist, even though in most cases one
can only present strong assumptions and not a formal proof of the intervention
of residues. In practice, it is indeed difficult to obtain a posteriori a
sample of food which at first sight was not doubtful. Most suspicions are
established as follows :

1° reaction of hypersensitivity, observed following food intake,

2° tests demonstrating that the individual is not allergic to the type of food
eaten but is to some drugs,

3° hence the possibility of the presence of those drugs in the food without
checking the hypothesis.

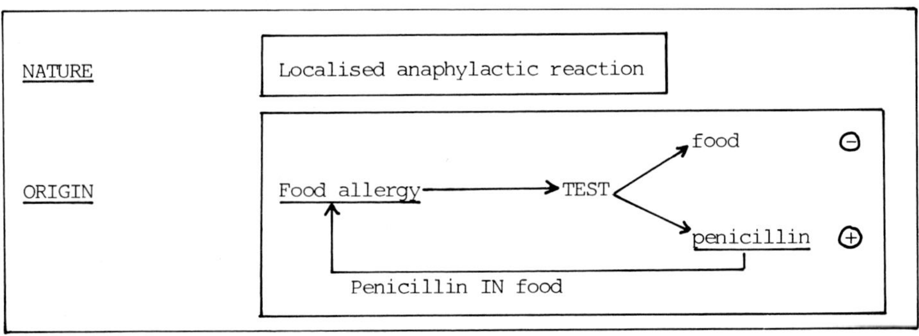

Fig. 8 : Food intolerance.

Some observations have been more complete. We have counted six, all
concerning penicillin. They have the advantage of proving the trigger mechanism of very low residual doses and constitute a quantitative approach of the
problem.

Four observations were made consecutive to consumption of milk in
which penicillin was identified and dosed at 4.2, 0.06 and 0.03 I.U./ml.
(Borrie, P. and Barrett, J., 1961; Erskine, D., 1958; Vickers, H.R., et al,
1958; Wicher, K. et al, 1969).

One observation concerned the intake by a butcher of meat from a pig
slaughtered in emergency 3 days after administration of penicillin. The
muscle contained 0.45 I.U. of penicillin per gram. This person had already
been treated with penicillin (Tscheuner, I., 1972).

Lastly, Wicher (1980) reports the extreme sensitivity of a person who
had already had reactions of intolerance to milk, but who presented once a

reaction of hypersensitivity after having drunk a fruit juice. This drink contained 0.3 I.U./ml of a "penicillin-like compound", a term which indicates the prudence and surprise caused by this result. One wonders if, in different circumstances, involving for instance milk or meat, the authors would have used the same wording. (Fig. 9).

| References | Quantities absorbed (I.U.) | | | Reactions |
	Concentration I.U./ml	Food	Total BP I.U.	
Wickers 1958	4	Milk : 1 l.	4,000	
Erskine 1958	0.06	Milk	Unknown	
Borrie 1961	0.03	Milk: 500 ml	15	Severe No effect dose: 15 IU
Wicher 1969	10	Milk: 240 ml	2,400	General pruritus, rash, headaches
Wicher 1980	0.6	Fruit juice 90 ml	54	General rash
Tscheuschner 1972	0.45 (UI/g)	Meat	Unknown	General rash.

These values are interesting to compare to the doses of penicillin mentioned :
- in allergy tests : 10,000 I.U. per os, 1 I.U. intradermally;
- in anaphylactic reactions : injection of 3×10^{-6} I.U.
- in lethal shocks : 1 I.U. after I.V. administration, 4,000 I.U. intradermally, 400,000 to 1 million I.U. orally.

These various elements enable us, at least in the current state of knowledge, to think that these highly antigenic substances present in food may cause allergenic reactions in some persons. Where residues are concerned, the doses are nevertheless very low, without relationship with the doses used for allergy tests or incriminated in lethal shocks.

3.   Control of the risk

a) Qualitative aspect : major allergenic substances

Intolerance of food origin described in human medicine are extremely numerous. Physicians however recognize that some substances are more sensitizing than others.
- Experimentally this phenomenon appears during detection tests, such as the Magnusson test where 5 classes are recognized. Penicillin always appears as

one of the most allergenic substances known (Magnusson, 1969). (Fig. 10)

I	VERY LOW	(0-8)	HEXACHLOROPHENE
II	LOW	(9-28)	BENZOCAINE
III	AVERAGE	(29-64)	SULFATHIAZOL
IV	HIGH	(65-80)	STREPTOMYCIN
V	VERY HIGH	(81-100)	PENICILLIN

Fig. 10 : Codification of the degree of allergenicity (Magnusson test).

- In practice the major allergenic substances incriminated in human medicine appear from various enquiries, such as that presented below, performed in France (Fig. 11) (Pradalier, 1980)

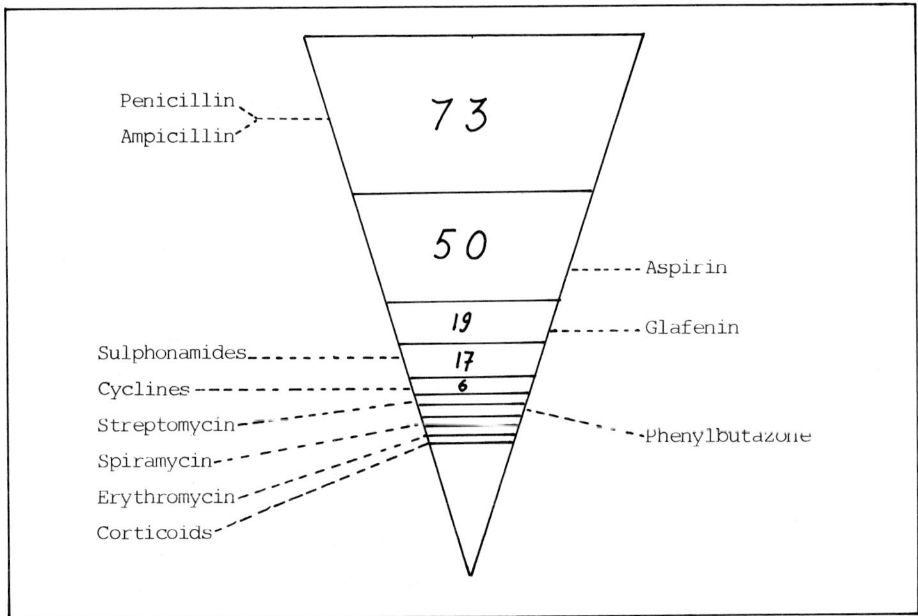

Fig. 11.

b) <u>Quantitative aspects</u>

    <u>Notion of threshold dose</u> : the research of "tolerance" to limit the risk requires that one first admits the notion of threshold dose in allergology. As for any type of immune response, it exists and can be demonstrated :

- fundamentally : interactions between antigens and antibodies,
- application : provocation tests, unsensitization therapy with "sufficient" dosages.

The quantitative evaluation remains difficult in view of the variations in existing factors :

- species : difficulty to experiment and extrapolate to man;
- individual : "hypersensitive" persons always represent a fraction of the population which it is difficult to appreciate and whose degree of hypersensitivity can hardly be defined. All preventive measure taken in human medicine accept the notion of "limited risk" without guaranteeing the total absence of risk.

In spite of this Lindmayr et al (1981) have not hesitated to research the consequences for known hypersensitive persons of the intake of meat of animal treated with BP. On 10 (volunteer) patients, only two had benign symptoms two hours after the meal (slight abdominal pruritus, sensations of anaesthesia at the level of the mouth). The authors conclude that the ingestion of such small doses of BP (6 to 12 I.U.) does not provoke allergenic reactions.

b) Tolerances

For benzylpenicillin, the examination of the various quantitative data permits us to propose a tolerated dose of 10 I.U.; i.e. for milk and dairy production (500 ml/day) a maximum concentration of 0.020 I.U./ml. This value is superior to the limit of the test currently used in France (Galesboot - 0.005 I.U./ml) and this analytical tolerance is thus satisfactory. For meat the same value can be proposed. (Fig. 12)

Oral route

Lethal shock	1,000,000 - 400,000 I.U.
Detection test	10,000
Desensitization	100 x n
Intolerance	4,000 - 15
No effect (on allergic subjects)	10

No effect : 10 I.U. Concentration that can be authorized   0.020 ul/ml

in 500 ml of milk (10 IU/500 ml)   0.010 ppm

(Galesboot, detection limit = 0.005 IU)

Other compounds : 0.010 - 0.100 ppm.

Fig. 12 : Benzylpenicillin tolerance.

154

For other veterinary drugs, we have no precise data and propose to reason by analogy, knowing that :

- immune response may now be interpreted on molecular bases. A demonstrated tolerance for BP could be transposed to a compound whose molecular weight is similar to BP. This is the case for many drugs : sulfonamides, chloramphenicol, tetracyclines, phenylbutazone ... This analogic approach gives a tolerance of 0.010 ppm (=0.020 I.U. of BP/g).

- this value could be weighted in relation with the allergenic potential of the considered compound.

In summary if it is certain that the consumption of food containing highly sensitizing substances, such as benzylpenicillin, could cause little severe allergic reaction in some hypersensitive individuals, the sensitizing action of residues appears improbable. The possibilities offered by current control enable us to limit their impact. It is nevertheless true that the use of such substances poses problems in animal production which are certainly unknown; the handling of medicated feeds or the preparation from premixes exposes individuals to a cutaneous contact with or the inhalation of products which must certainly be immunogenic in these conditions.

References

Borrie, P., Barrett, J., 1961. Dermatitis caused by Penicillin in bulk milk supplies. Br. Med. J., Nov., 1267.

Bundgaard, H., 1983. Chemical and Pharmaceutical aspects of drug allergy. Allergic reactions to drugs. de Weck and Bundgaard, Ed. Springer-Verlag, 37-74.

Erskine, D., 1958. Dermatitis caused by Penicillin in milk. The Lancet, 1, 438.

Gounelle, H., Szakvary, A., 1966. Antibiotiques et aliments. 1. Les accidents allergiques liés aux résidus. Bull. Acad. Nat. Med., 50 : 76-82.

Idsoe, O., Guthe, T., Willcox, R.R., de Weck, A.L., 1968. Nature et importance des réactions secondaires à la Pénicilline, compte tenu notamment des cas mortels par choc anaphylactique. Biol. Med., 57 : 402-448.

Lindemayr, H., Knobler, R., Kraft, D., and Baumgartner, W., 1981. Challenge of Penicillin-contaminated meat. Allergy, 36 : 471-478.

Magnusson, B., Kligman, A.M., 1969. The identification of contact allergens by animal assay, the guinea pig maximization test. J. Invest. Dermatol., 52, 268-276.

Ponvert, C., Saurat, J.H., Tilles G., Galoppin, L., Paupe, J., 1978. Test de migration leucocytaire et allergénicité d'un médicament. Thérapie, 33 : 455-461.

Pradalier, A., Dry, J., Luce, H., 1980. Réflexion sur l'allergie médicamenteuse. Concours Médical, 40 : 5993-6011.

Tscheuschner, I., 1972. Anaphylaktische reaktion auf Penicillin nach Genuss von Schweinefleisch. Z. Haut-Geschl. Kr., 47 : 591-592.

Vickers, H.R., Bagratuni, L., Alexander, S., 1958. Dermatitis caused by Penicillin present in milk. The Lancet, 351.

Wicher, K., and Reisman, M.R.E., 1980. Anaphylactic reaction to Penicillin (or penicillin-like substances) in a soft drink. J. Allergy Clin. Immunol., 66 : 155(157.

Wicher, K., Reisman, R.E., Arbesman, C.E., 1969. Allergic reaction to Penicillin present in milk. J.A.M.A., 208 : 143-145.

156

RESUME

RISQUES ALLERGENES LIES AUX RESIDUS DE MEDICAMENTS VETERINAIRES

Viviane BURGAT-SACAZE,
Laboratoire de Toxicologie Biochimique et Métabolique (INRA)
Ecole Nationale Vétérinaire de Toulouse,
F-31076 Toulouse Cedex, France

Les résidus de médicaments vétérinaires sont parfois incriminés dans des intolérances d'origine alimentaire observées chez l'homme. La complexité des phénomènes et les molécules impliquées (le plus souvent des antibiotiques) ont souvent amené les hygiénistes à recommander une grande prudence en la matière, sans cependant apprécier le risque réel en cause. Notre propos aura trait à cette appréciation qui nous a conduit, d'une part, à rechercher la nature du risque encouru et, d'autre part, à proposer une méthode d'évaluation de ce risque.

Les caractéristiques des réactions d'hypersensibilité sont à l'heure actuelle bien codifiées : les intolérances d'origine alimentaire appartiennent essentiellement aux phénomènes du type I (hypersensibilité immédiate). Elles se différencient des réactions allergiques médicamenteuses iatrogènes.

Le bilan des accidents attribués aux résidus de médicaments vétérinaires fait, par ailleurs, apparaître que ces manifestations sont rares et jugées médicalement peu graves (urticaire, oedèmes localisés).

En fonction de ces diverses données théoriques et pratiques, les résidus peuvent être incriminés dans des réactions déclenchantes. L'effet sensibilisant paraît peut probable.

Un deuxième aspect concerne la prévention de ces effets indésirables : une approche qualitative et quantitative du risque peut être effectuée en recherchant les relations dose-effet en la matière, qui permettent :

- de préciser la nature des principaux allergènes en cause,

- de proposer des doses résiduelles sans effet en fonction de l'allergénicité des composés.

ZUSAMMENFASSUNG

ALLERGIE : MECHANISMEN UND URSACHEN

Viviane Burgat-Sacaze,

Laboratoire de Toxicologie Biochimique et Métabolique (INRA)

Ecole Nationale Vétérinaire de Toulouse,

F-31076 Toulouse Cedex, Frankreich.

Rückstände von Tierarzneimitteln wurden in mehreren Fällen bei Unverträglichkeiten des Menschen gegenüber Nahrungsmitteln als Schuldige bezeichnet. Die Komplexität dieser Erscheinungen und die implizierten Stoffe, meistens Antibiotika, haben oft Hygieniker veranlaßt, eine starke Zurückhaltung zu empfehlen, ohne jedoch das reelle Risiko zu bewerten. Der vorliegende Beitrag nimmt Bezug auf diese Bewertung, die uns auf der einen Seite dazu geführt hat, die Natur des Risikos zu erforschen, und uns auf der anderen Seite veranlaßt, ein Verfahren für die Bewertung dieses Risikos vorzuschlagen.

Die Charakteristika der Überempfindlichkeitsreaktionen sind heutzutage gut bekannt : die Unverträglichkeit im Hinblick auf Nahrungsmittel gehört im Wesentlichen zu den Erscheinungen des Typs I (Überempfindlichkeit, die sich sofort äußert). Sie unterscheiden sich von den iatrogenen, auf Medikamenten beruhenden allergischen Reaktionen.

Eine Bilanz der Fälle, die auf Rückstände von Tierarzneimitteln zurückgeführt werden, zeigt übrigens, daß diese Manifestationen nur selten auftreten und von medizinischer Seite her als wenig schwerwiegend eingestuft werden (Nesselfieber, lokalisierte Ödeme).

Auf Grund dieser unterschiedlichen theoretischen und praktischen Gegebenheiten lassen sich die Rückstände als verantwortlich für Auslösereaktionen bezeichnen. Eine sensibilisierende Wirkung erscheint wenig wahrscheinlich.

Ein zweiter Aspekt betrifft die Prävention solcher unerwünschter Nebeneffekte : eine qualitative und quantitative Risikoannäherung läßt sich durchführen, indem man das Verhältnis zwischen Dosis und Wirkung untersucht. Dies erlaubt uns :
- die Natur der wichtigsten Allergene zu präzisieren
- im Hinblick auf die allergene Wirkung der Zusammensetzungen verträgliche Rückstandsdosen zu definieren.

<u>ANTIBACTERIALS AND RESISTANCE TO THEM IN</u>
<u>THE TREATMENT AND CONTROL OF ANIMAL DISEASE</u>

D.J. Taylor,
Department of Veterinary Pathology,
University of Glasgow Veterinary School,
Bearsden Road, Bearsden, Glasgow G61 1QH.

<u>Introduction</u>

Antimicrobials belonging to at least ten distinct and unrelated groups
of compound are available to veterinarians for the treatment and control of
animal disease in most countries of Western Europe. Some of these antimicrobials have been used for 40 years and others for as little as two or three.
All play a part in the treatment of microbial disease in animals and in aiding
the veterinarian in his four main duties to his patients and his clients.
These are :
1. to eliminate pain and suffering in animals under his care as rapidly as
    possible;
2. to restore the animal to full productivity within the shortest possible
    period of time;
3. to reduce or eliminate any risk posed by the animal disease to the human
    population;
4. to reduce or eliminate any risk the disease may pose to other animals on
    the infected premises or to the animal population at large.
In any discussion of the use of these compounds, these duties must be
borne in mind.

Antimicrobials form only one of a number of therapeutic and preventive
measures which may be used to treat and control disease caused in animals by
microorganisms susceptible to antimicrobials. Many of these other methods can
be applied to infectious disease generally and include immunotherapy, isolation, disinfection, vaccination, varying forms of supportive therapy including
rehydration and, finally, management changes. Few of them can be used in the
treatment of acute bacterial disease in animals in the same way as antimicrobials.
Similar methods can be used for the control of animal diseases but
there is one major difference between the treatment and control of animal

disease and that of human disease. This is that the animal may be slaughtered if suffering from severe pain or if it is unlikely to recover and in certain diseases, such as brucellosis and tuberculosis, slaughter of infected animals and safe disposal of their carcases is required by law.

In considering the subject of antibacterials and resistance to them in the treatment and control of animal disease it may be helpful to consider the way in which antimicrobials are used in the treatment and control of animals disease in Europe and then to consider the problems which arise from these treatments.

1.   The use of antimicrobials in the treatment of animal disease

a) Treatment : In its simplest sense, this is carried out by a veterinarian upon being presented with a sick animal by a client. The antimicrobial chosen, the route of administration and the duration of the course of treatment will depend upon the type of infection diagnosed by its clinical signs or after the laboratory examination of samples. This type of diagnosis and treatment is most common in household animals such as dogs and cats or cage birds, horses and individual cows or calves. In other animals such as pigs, sheep or chickens, the group or herd may be treated as a group in the feed or drinking water.

In most cases the administration of second or subsequent doses of antimicrobial is left to the owner (dogs, cats and horses) or the owner and his employees (farm animals). This repeat treatment is therefore outside the control of any professionally qualified person.

Treatment for bacterial diseases is often accompanied by supportive measures such as isolation, warmth, special food or rehydration and these procedures may be essential for successful recovery.

b) Prevention : Antimicrobials are often given to prevent the likely occurrence of infection at times of exposure. An example of this is the use of local antimicrobial dusting or parenteral treatment at the time of wound treatment or after surgical procedures. In a similar way, disease which occurs in animals after weaning or after movement can be reduced in severity or eliminated by giving antimicrobials appropriate to the disease which will occur in the group. An example is postweaning diarrhoea in pigs associated with E. coli of pathogenic serotypes.

Lower levels of antimicrobial may also be included in the rations of such exposed animals to prevent infection. Such treatments are generally necessary for only a short portion of the animal's life.

c) Strategic treatments : Bacterial disease may be controlled or eliminated from a unit by whole herd treatment against a specific organism.

This is most frequently practised in pig or poultry enterprises. Here
mycoplasma infections have been eliminated by egg dipping followed by testing
for carriers and their removal. Similar programmes have been used to
eliminate swine dysentery, enzootic pneumonia and atrophic rhinitis from pig
herds. In many pig and poultry enterprises where an all-in, all-out system is
practised, animals introduced to a clean unit are treated to eliminate
infection brought with them and normally remain free from these infections for
long period.

d) Growth permitters : Certain antimicrobials have been used
throughout Western Europe for inclusion in the rations of growing pigs,
poultry and cattle to reduce subclinical disease which results in reduced
growth. Few of these antimicrobials have any influence on identifiable
clinical disease. They are relatively freely available to farmers through
feed merchants and agricultural pharmacists and they are included only at low
levels. Strict criteria are applied to the release of compounds for this use
within the European Community and those in use do not in general affect the
use of any therapeutic antimicrobials.

2.   The results of antimicrobial treatment in animal disease

a) Successful treatment : In spite of concern from some quarters
about antimicrobial resistance, it should be clearly understood that the vast
majority of antimicrobial treatments in cases of bacterial disease in animals
have a successful outcome.

Many important infectious diseases of animals and a large number of
local lesions are caused by organisms that are sensitive to penicillin. The
local lesions include abscesses such as those which result from fighting
between cats and which yield streptococci or pasteurellae in culture, i.e.
antimicrobial sensitive organisms. Figures from the U.K. indicate that, of
the cases of farm animal disease submitted to the Government Veterinary
Investigation Centres in 1982, infections due to organisms that are always or
often sensitive to antimicrobials formed the majority (Table 1).

These figures are merely a guide. The vast majority of cases which do
not present any problems in treatment never reach the Veterinary Investigation
Centres at all. The 'cases' themselves may, however, indicate disease in
tens, hundreds or, in the case of poultry, thousands of individuals.

In all the animal species, bacteria such as E. coli, Salmonella spp.,
Staphylococcus spp., Mycoplasma spp., Pasteurella spp. and Treponema spp. are
sensitive to several different antimicrobials. Resistant strains or isolates
have been recorded in all of these groups and the importance of these will be
considered below.

TABLE 1

Cases* of farm animal disease diagnosed as being caused by bacteria in Government Veterinary Investigation Centres U.K. 1982 analysed by probable response to antimicrobials.

Species	Cattle	Sheep	Pigs	Poultry
Total number of cases	101,962	24,328	15,530	9,421
Total bacterial disease	16,942	3,983	4,091	1,592
Bacterial disease sensitive[+] to				
penicillin - number	5,286	887	750	54
percentage	31	22	18	3
Bacterial disease due to almost wholly resistant organisms[x]				
- number	859	–	–	103
percentage	5	–	–	7

* 'Cases' may represent individual animals or groups.
+ Staphylococci are considered as not penicillin sensitive in this context.
x Mycobacteria, Brucella and Pseudomonas are considered almost wholly resistant.
Source : Veterinary Investigation Diagnosis Analysis II 1982, M.A.F.F.

b) <u>Unsuccessful treatment</u> : Antimicrobial treatments may have an unsuccessful outcome in animal disease for a number of reasons. One of these is antimicrobial resistance and it will be discussed in detail below. The importance of resistance as the cause of failure of treatment is not entirely clear and the other reasons for failure may be equally important. They will be considered briefly and then antimicrobial resistance will be discussed in greater detail.

Reasons for the failure of antimicrobial treatment include :

i) Incorrect diagnosis of a disease as bacterial. This may occur when diseases caused by agents other than bacteria have the same clinical signs as bacterial disease. This is especially common in enteritis in which more than one viral parasitic or bacterial species may be involved. In such cases the veterinarian often treats the animal immediately as if the condition were bacterial and reconsiders his treatment if it fails or when laboratory examinations are reported after 3 to 5 days.

ii) Incorrect identification of the bacteria present. Bacterial diseases may present with similar clinical signs. <u>E. coli</u> and <u>Pseudomonas</u> mastitis may present in a similar way but treatment adequate for the former may have no effect on infection with the latter. Laboratory examinations and antimicrobial sensitivity determinations usually reveal an appropriate antimicrobial.

iii) Treatment too late to affect the outcome. The level of supervision of individual animals is often poor, both in extensive husbandry systems such as rough grazings and in intensive husbandry systems. In both situations animals may become ill and become moribund before being noticed. This is particularly common during the neonatal period when mortality in domestic species may vary from 5 % (well-run intensive piggery) to 20 % or more (extensive sheep husbandry).

In addition to this, some diseases such as anthrax in cattle, streptococcal meningitis and pleuropneumonia in pigs and erysipelas infection in fowls and turkeys may be of such rapid onset (6 to 12 hours) that treatment is impossible although the causal bacteria are fully sensitive to antimicrobials. Toxins produced by organisms such as <u>Clostridium</u> <u>tetani</u> and <u>E. coli</u> (in Oedema disease) may fix to tissue and the clinical signs may be irreversible even when the organism has been eliminated.

iv)   Treatment given in inadequate quantity or by an inappropriate route. In many cases where therapy of bacterial disease is attempted in domestic animals the amount of antimicrobial given may be inappropriate. It is uncommon for larger animals such as bulls and horses to be given the correct dose of antimicrobial in mg/kg. Often, as a strong or dangerous animal recovers in extensive husbandry, the animal cannot be recaptured for a second dose and the owner decides that the animal is better and fails to give a second treatment or to call the veterinarian to do so.

Treatments in water and feed may be curtailed and inadequate because of cost, because of failure to incorporate the drug at the correct level in feed or water, or because the medicated feed is being fed as only part of a ration e.g. in swill-fed or whey-fed pigs. At times the feed intake per kg liveweight is unlikely to allow the intake of a therapeutic dose as in the dry sow and the fattening pig on a restricted ration.

Finally, sick animals may not eat and may therefore fail to ingest adequate quantities of food to allow therapy as in pneumonia.

v)   Inadequate supportive measures. In some disease syndromes such as bacterial meningitis and enteritis, the organism may be killed or inactivated by the antimicrobial but in the absence of nursing care the animal may die from dehydration, chilling or starvation. When fed carefully, kept warm and quiet and rehydrated, affected animals may recover completely although after antimicrobial treatment alone they may die.

vi)   Failure to prevent reinfection in susceptible groups of animals. Too short a period of treatment, inadequate appreciation of the numbers exposed to infection (all those in airspace or drainage contact in respiratory and enteric diseases respectively) and failure to disinfect a unit in which pathogens survive can lead to reinfection. In groups of animals the identity of the clinically affected animal is rarely recorded and it appears as if 'the disease' has recurred. This is often attributed to antimicrobial resistance.

vii)   Antimicrobial resistance. Antimicrobial resistance can be considered to be the cause of failure of treatment only after considering all the above points fully and demonstrating that tne actual causal agent of the disease is resistant to the antimicrobial being used.

The importance of antimicrobial resistance in animal disease varies according to the species and to whether the individual or the population is at risk. The individual may succumb to disease before an appropriate antimicrobial is found and if supportive therapy is inadequate, but the population on an individual unit is likely to survive. The disease can be controlled with an appropriate antimicrobial for clinical cases and hygiene, isolation and vaccination may be used to prevent spread within the unit and spread to other units. Where movement of animals between units is controlled the majority of bacterial diseases cannot spread to adjacent units and the remainder of the national herd is not at risk.

It may be of interest to consider here some diseases or syndromes in which antimicrobial use has resulted in resistance and the measures used to treat and control these diseases.

<u>Mastitis</u> : Bovine mastitis is an important syndrome and accounted for almost 11 % of all cases of bovine disease submitted to Veterinary Investigation Centres in the U.K. in 1982 and 65% of all cases of bovine bacterial disease investigated. The bacteria isolated were as per Table 2 (with common antimicrobial sensitivities in brackets).

TABLE 2

Totals of bacteria isolated from bovine mastitis by Veterinary Investigation Centres 1982.

	Number of cases	Approximate sensitivity to penicillin		Significant resistance to common antimicrobials	
Totals	10,977				10 %
C. bovis	172	S	(90 %)	St.	(40 %)
C. pyogenes	543	S	(90 %)	PN	(40 %)
S. agalactiae	367	S	(100 %)	St.	(100 %)
S. dysgalactiae	1,139	S	(100 %)	St.	(22 %)
S. uberis	2,124	S	(100 %)	St.	(82 %)
Other streptococci	215	S	(100 %)	St.	
S. aureus	1,877	R	(60 %)	St.	(20 %)
				PN	(60 %)
Other staphylococci	541	S	(80 %)	PN	(16 %)
E. coli	2,644	R	(100 %)	SXT	(28 %)
Klebsiella	101	R	(100 %)	SXT	(30 %)
Pseudomonas	267	R	(100 %)	St, PN, SXT, etc.	
Mycoplasma spp.	10	R	(100 %)	St, PN, SXT, etc.	

S = sensitive   St. = Streptomycin   SXT = trimethoprim/
R = resistant   PN = Ampicillin   sulfonamide

Source : Veterinary Investigation Diagnosis Analysis II, 1982, M.A.F.F.

E. coli is largely environmental in origin and is largely sensitive to commonly used intramammary antimicrobials other than penicillin.

Current figures for the National (U.K.) dairy herd suggest that, although the individual yield per cow is rising as is the number of cows per herd, the cell count (a non-specific indicator of mastitis) and the percentage of penicillin resistant staphylococci is falling (Table 3).

TABLE 3

Changes in the dairy herd of England and Wales and levels of yield, mastitis and resistance in staphylococcal isolates.

Year	Average herd size	Average yield	Average cell count	% Staphylococcal isolates penicillin sensitive
1979[+]	56	4,680	478,000	66.5
1980	58	4,715	469,000	62.0
1981	63	4,810	464,000	46.5*
1982	65	4,800	427,000	–
1983	–	5,085	390,000*	–

Figures from U.K. Dairy Facts and Figures 1983 and * from J. Booth, M.M.B. personal communication.

[+] Some recording years do not coincide with calendar years.

These figures are of particular interest because 34 % (by value) of all anti-microbials sold in the U.K. are intramammary preparations.

This decline in the resistance to penicillin has been attributed to the increased use of dry cow therapy, increased hygiene and more rational culling. All these factors have reduced the number of treatments per lactating cow.

Enterobacterial infections : The Enterobacteriaceae are particularly likely to develop antimicrobial resistance, but in most cases this is not complete and alternative methods of therapy and hygiene aid in their treatment and control. E. coli and the Salmonellae are the members of this family most likely to cause disease in animals. E. coli disease will be considered here and the salmonella infections below under the section of human health hazards.

The majority of E. coli infections in calves, pigs, lambs and other species are enteric in nature, although in chicken septicaemic infection is more important. Resistance may occur to antimicrobials used for therapy but this need not be a problem as a few isolates from enteric disease are completely resistant to all antimicrobials available. Electrolyte therapy is usually adequate to overcome the acute effects of disease and the vaccines and hygienic measures available mean that long term control need not be a problem in

these species. The situation is more difficult in infections in poultry, but
even here hygiene is more important in control and can prevent or reduce infection.

The rarity with which E. coli becomes totally resistant is shown in
Table 4 in which the antimicrobial sensitivity of 100 E. coli isolates from
dogs is shown. The organisms were isolated from cases of enteritis, cystitis
and wound infection and were considered to be causal or contributory in these
syndromes. In many cases treatment had been given before referral. Only 2 %
were completely resistant to the antimicrobials listed and they were sensitive
to substances such as colistin, gentamycin, apramycin or spectinomycin when
tested.

TABLE 4

Sensitivity to antimicrobials of E. coli isolates from canine disease 1983,
Glasgow University Veterinary School.

Antimicrobial	Number sensitive (percentage)
Ampicillin	59
Chloramphenicol	85
Furazolidone	68
Neomycin	59
Oxytetracycline	35
Streptomycin	48
Sulphafurazole	12
Trimethoprim sulphonamide	58

Respiratory disease : Of more importance to the veterinarian is the
resistance of respiratory pathogens such as Pasteurella spp., Haemophilus spp.
and Bordetella bronchiseptica to oxytetracycline. Isolation and husbandry
methods of disease control are more difficult to carry out with respiratory
pathogens, particularly in animals such as sheep and cattle. Trimethoprim
sulphonamide and penicillins are almost always effective against the
pasteurellae and vaccines are becoming more readily available. Trimethoprim
sulphonamide is less effective against Bordetella in all species, but vaccines
are now available. Only Haemophilus spp. still causes problems in disease and
cannot be prevented by vaccination in the U.K. In other countries of Western
Europe vaccines are available and control may be easier.

Swine dysentery : Finally, in swine dysentery, the causal bacterium,
the spirochaete Treponema hyodysenteriae is almost completely resistant to
macrolides because of widespread therapeutic use of tylosin. Resistance to
lincomycin has occurred frequently but resistance has not yet been recorded
with tiamulin. Some resistance to dimetridazole and ronidazole appears to
occur, but this seems not to be maintained. Finally, some agents such as

monensin and carbadox appear incapable of inducing sustained resistance. At
least one of the available agents can always be used to control the disease in
spite of failure to respond to one or more of the agents available.
Parenteral administration of these drugs is often effective when they are
ineffective per os.

Swine dysentery is, therefore, a typical example of a bacterial
disease of animals in Western Europe. Some resistance does occur and, indeed,
may be expected, but the choice of drugs is sufficiently wide and general
methods of control sufficiently well understood that direct economic loss from
clinical disease and death are the exception rather than the rule.

3.   Antimicrobial use in animal disease as a threat to human health

a) From resistant organisms derived from animal disease : Many
diseases of animals are a threat to human health, but few of these are
affected by the use of antimicrobials in the treatment of animal disease.
Tetanus, Leptospiral, Anthrax and Streptococcus suis infections are all fully
sensitive to penicillin, while the chlamydia of psittacosis and ornithosis and
the rickettsia of Q fever appear to respond fully to tetracycline therapy in
man. Mycobacterial infections caused by M. bovis, brucellosis, and Glanders
(Pseudomonas mallei) have all been controlled by non-antimicrobial means.
These are essentially based upon the detection of infected animals by the
clinical signs of the disease or by a blood or skin test which reveals carrier
animals which are then slaughtered. The major source of infectin for man in
tuberculosis and brucellosis was milk and, pending the eradication of these
diseases, this route of infection was eliminated by pasteurisation.

The widespread use of pasteurisation (compulsory in many European
countries and soon to be universally so in the EC) eliminates any possibility
of the transmission of bacteria to the general population in milk. Thus only
those consuming unpasteurised milk or milk products will be at risk from milk-
borne agents such as M. bovis, Q fever, B. abortus, B. melitensis, L. hardjo,
Salmonella spp. and Campylobacter spp. These are usually the farm workers or
other elements of the rural population in intimate contact with the animals.
All mastitis pathogens are eliminated in this way and any antibacterial resistance in them is therefore irrelevant except to those drinking unpasteurised
milk. Even here there is no evidence that E. coli from milk or staphylococci
(Lacey, 1981) represent a threat to human health.

There appears to be no evidence that E. coli of animal origin are
involved in human disease. The enteropathogenic strains of animal origin
appear to have different adhesive determinants from those which cause human
disease.

It is theoretically possible that multiply-resistant E. coli of animal origin could colonise the human gut and transfer the resistance to resident E. coli. Unfortunately no study has been carried out on multiply-resistant E. coli from human disease to see whether the plasmids they contain might be of animal origin. All the evidence from medical data seems to indicated that E. coli from non-enteric sites are more resistant when isolated from hospital patients than when isolated from those from general practice. A recent study from Ruchill Hospital, Glasgow (R. Campbell Tait, unpublished data) confirms this (Table 5).

TABLE 5

Antimicrobial sensitivity of Urinary E. coli from both hospital and outpatient sources.

	Inpatient	Outpatient
Number	89	164
% sensitive to Amoxycillin	42	58*
Cephalosporin	80	91
Tetracycline	68	69
Nalidixic acid	94	96
Nitrofurantoin	98	100
Sulphafurazole	51	51
Trimethoprim	78	88*
Cotrimoxazole	80	89*
Augmentin	93	98

* Antimicrobial sensitivity approximately 10 % lower in hospital isolates than in those from general practice.

Source : R. Campbell Tait, Ruchill Hospital, Glasgow, unpublished data.

Any E. coli of animal origin that did cause enteric disease would be extremely unlikely to be treated with antimicrobials and resistance is therefore likely to be irrelevant, and the evidence cited above suggests that the medical use of antimicrobials is more important than that in animals in E. coli infections in man.

Two organisms which may become resistant are frequently transmitted to man from animals. Neither may actually be treated in the animal of origin as a cause of clinical disease. These are the salmonellae and the campylobacters. Campylobacters cause enteric disease and studies suggest that treatment may only reduce the course of the clinical disease by one day. Infection rarely threatens life and is usually enteric. C. jejuni or C. coli infections are rarely treated specifically in animals or poultry and any antimicrobial resistance which is present has usually been acquired as a result of exposure to therapeutic antimicrobials given for other reasons. The vast majority of campylobacters isolated from dogs, cats, cattle and poultry are sensitive _in vitro_ to neomycin, metronidazole and erythromycin. It is possible that the use of neomycin and dimetridazole or other nitroimidazoles (in therapy) or tylosin and other macrolides (in therapy and as growth permitters) may affect this, but even if resistance to erythromycin were to become common, the value of treatment in man is, in any case, finely balanced.

Salmonellae also cause disease, both in man and in domestic animals. Man can become infected from animals and _vice versa_. Most infections are enteric and in man are untreated except, perhaps symptomatically. The most important human pathogens, _S. typhi_, _S. paratyphi_ A and B, are not found in animals and antimicrobial treatment in animals is unlikely to affect them. Most infections of animal origin which require treatment in man are caused by _S. typhimurium_. This serotype is currently less sensitive than other serotypes to antimicrobials, largely because of the spread throughout the UK of a single phage type, DT204 and its relatives which are multiply-resistant. _S. typhimurium_ occurs in all species, but DT204 is most common in calves and spreads by means of calf movement. Isolates of this phage type are multiply-drug-resistant but are all sensitive _in vitro_ to furazolidone. Thus short term treatment need not be a problem. Isolation of calves with resistant or sensitive strains of _S. typhimurium_ results in the elimination of the organism within six weeks, so long term control within a unit is straightforward. If calf movements are controlled the spread of the infection can be controlled on a national basis as happened with a previous multiply-resistant phage type (DT29) which died out after the dealer, whose premises were infected, went out of business.

Sojka _et al_ (1984) have provided data on the sensitivity to antimicrobials of 12,903 salmonella isolates from animals in the UK and have shown (Table 6) that many of them are fully sensitive to antimicrobials and therefore present no problem in terms of resistance. It is notable that serotypes other than _S. typhimurium_ are more sensitive than that serotype. As many salmonella serotypes cause inapparent or undiagnosed infections in animals, it is unlikely that treatment of the disease in animals radically affects the transmission. In some situations it may be of positive benefit as in calf salmonellosis where the organism is reduced in number and may therefore be less likely to spread to man. The total pasteurisation of milk is likely to reduce the transmission of salmonellae from cattle to man quite markedly and to restrict infection to those working on farms or in direct contact with cattle or their products.

In general, the bacteria transmitted from animals to man are relatively unaffected by antimicrobial treatment in animals. The exception is

TABLE 6

Summary of some features of antimicrobial sensitivity in Salmonella isolates
from animal source 1979-81

Year	Total sensitive strains %	S. typhi-murium %	S. dublin %	Other serotypes %
1979	22.9	16.2	20.6	29.8
1980	31.4	17.4	37.5	41.2
1981	10.9	6.1	7.6	16.0

Source : Sojka et al (1984).

perhaps, the salmonellae, but Lacey (1984) has reviewed the current situation
in a paper entitled "Does the use of chloramphenicol in animals jeopardise the
treatment of human infections". He considered that chloramphenicol was not
now the antimicrobial of first choice but that trimethoprim was. It is clear
from animal data (Sojka et al., 1984) that trimethoprim resistance was increasing (through the spread of S. typhimurium phage types 204 and 193) particularly in isolates from cattle but this possible threat is easily controlled by animal movement control, hygiene and isolation given the political
will. No hygienic measures are ever going to prevent the spread of salmonellae to man, but measures such as slaughterhouse and kitchen hygiene, pasteurisation of milk and the proper cooking of food would do much to reduce the
level of human infection. Antimicrobials play little part in this control as
summarised by Lord Swann recently in the House of Lords (Hansard 25th January
1984). He considered that the control of the use of antimicrobials in feedstuffs which he had recommended had failed because of the use of antimicro
bials in therapy which had generated drug resistance in bacteria. He considered the control of antimicrobial resistance in bacteria a lost cause as the
only controls which would be effective would hamstring both human and veterinary medicine. He advocated a return to old-fashioned methods such as
isolation, disinfection, etc.

This view has been discussed above and examples of the role of antimicrobials and other methods in the control of animal disease has been fully
discussed.

b) Prevention of residues of antimicrobials from reaching the consumer

All antimicrobial preparations are marketed with a recommended withdrawal
time. The veterinarian normally informs the farmer of the correct withdrawal
period for the antimicrobial and does not send animals for emergency slaughter
without declaring the treatments which they have received. This should dis-
charge the veterinarian's obligation to the purity of meat and milk presented
for sale. In the dairy field it is normal custom to test milk for residues of
antimicrobials weekly or more in the U.K. Failures are punished by financial

sanctions and the levels of failure are low. In the case of meat animals some treatments may still be continuing at slaughter as a result of inadequate supervision by the owner, but this cannot be controlled by the veterinarian at farm level. With adequate veterinary advice, good husbandry and hygiene practices, treatments should be minimal in animals of slaughter age and the residues from earlier treatments should be long gone at slaughter. The fate of residues of antimicrobials in milk and meat has been reviewed recently by Walton (1983) who recommended a simple, non-specific testing regime which would detect antimicrobials present above levels which are agreed to affect human health.

Conclusion

1.   Antimicrobials are of value in the treatment of animal disease and help reduce suffering and aid in the rapid return of the affected animal to full productivity.

2.   Strategic medication at therapeutic levels coupled with isolation, cleaning and vaccination can markedly reduce the overall level of disease in a herd, eliminate some pathogens and reduce the total number of antimicrobial treatments required.

3.   Resistance to antimicrobials arises largely as a result of therapy but the vast majority of animal infections are sensitive to the antimicrobials currently available to the veterinarian. With an increasing number of new compounds being presented for development, resistance is less likely to be a problem in future.

4.   When veterinarians are in charge of diagnosis and prescription (as in the U.K.) any problems with antimicrobial resistance which arise can be dealt with and overcome by the use of alternative or supplementary methods. When therapeutic antimicrobials are widely available to clients and owners, the reasons for failure to respond to antimicrobial treatment may not be fully appreciated and inappropriate therapy may be continued.

5.   Antimicrobial resistance plays little part in the national control of microbial disease. Most control schemes rely on the slaughter of infected animals, the safe disposal of their carcases and disinfection of the unit and the control of animal movement. Only in the tetracycline treatment of imported parrots is antimicrobial used at a national level.

6.   The use of antimicrobials in animals poses little additional risk to human health from resistant organisms. Antimicrobial resistant organisms are no more pathogenic than sensitive ones. The greatest threat comes from antimicrobial resistance in salmonellae and movement control, isolation, vaccination, kitchen hygiene, adequate cooking and the pasteurisation of milk could

all reduce the transfer of these salmonellae to the human population. The
majority of human infections with salmonellae of animal origin are enteric and
are not treated with antimicrobials.

7.   When properly prescribed and used as directed, antimicrobial residues
in milk and meat are minimal. Testing for levels of antimicrobial agreed to
be harmful linked to financial penalties would ensure that the chance of
contamination would be further reduced.

<u>References</u>

Lacey, R.W., 1981. Are resistant bacteria from animals and poultry an
     important threat to the treatment of human infections ? In : Ten Years
     from Swann, A.B.P.I. London.
Lacey, R.W., 1984. Does the use of chloramphenicol in animals jeopardise the
     treatment of human infections ? Vet. Rec. 114 : 6-8.
Sojka, W.J., Wray, C., and MacLaren, I., 1984. A survey of drug resistance in
     Salmonellae isolated in England and Wales from 1979-1981. Br. Vet. J.
     (in press).
Walton, J.R., 1983. Antibiotics, Animals, Meat and Milk. Zbl. vet. Med. A.
     30 : 81-92.

RESUME

ANTIBACTERIENS ET RESISTANCE DANS LE TRAITEMENT ET LE CONTROLE DES MALADIES DES ANIMAUX

David J. TAYLOR,
Department of Veterinary Pathology,
University of Glasgow Veterinary School,
Bearsden Road, Bearsden, Glasgow G61 1QH.

Les antibactériens sont largement utilisés dans le traitement des maladies des animaux pour réduire douleur et souffrance, prévenir les mortalités, guérir les animaux cliniquement affectés et les ramener à leur niveau de rendement dans les plus brefs délais. Un nombre d'antibactériens non apparentés sont à la disposition des vétérinaires pour le traitement des animaux dans la plupart des pays d'Europe Occidentale. Ces antibactériens peuvent être utilisés individuellement chez les animaux ou servir dans des traitements de masse, et sont administrés soit par piqûre, soit dans l'eau de boisson, soit dans la nourriture à des doses thérapeutiques, ou encore en application locale sur la peau. Ils peuvent être utilisés également pour prévenir des maladies qui autrement se développeraient ou pour éliminer une infection d'un groupe par traitement stratégique. Certains antibactériens quelque peu apparentés à ceux utilisés thérapeutiquement sont inclus dans les aliments comme promoteurs de performance.

La plupart des traitements antibiotiques sont couronnés de succès et le nombre d'infections importantes qui restent sensibles à l'action de la pénicilline est considérable. Les chiffres de maladies diagnostiquées par les Laboratoires Vétérinaires Nationaux de Grande-Bretagne indiquent que 18 à 30 % des infections microbiennes diagnostiquées chez les bovins, les moutons et les porcs sont causées par des organismes sensibles à la pénicilline et que seulement 5 à 7 % de toutes les infections microbiennes sont associées à des organismes résistants comme les <u>Pseudomonas</u> spp., Mycobacteria et Brucella.

L'échec d'un traitement peut avoir de nombreuses causes, notamment la mauvaise identification d'une maladie comme étant non-microbienne ou la mauvaise identification de l'agent causal, un traitement commencé trop tard, une médication insuffisante, une thérapie supportive insuffisante ou de ne pas avoir pris les mesures pour empêcher la réinfection. La résistance aux antibactériens existe, mais peu de syndromes sont réellement incontrôlables avec le choix d'antibiotiques disponibles actuellement. A long terme, une bonne gestion et des politiques de prévention et de vaccination permettent de contrôler les infections. Quelques exemples de syndromes tels que la mammite bovine, l'entérite due à E. coli, les maladies respiratoires et la dysenterie porcine sont discutés et les moyens de contourner les problèmes de résistance sont présentés.

Finalement, le risque pour l'homme, suite au traitement des animaux avec des antibiotiques, est considéré. Les organismes, tels que Leptospira hardjo, Bacillus anthracis et Chlamydia psittaci sont tous sensibles aux antibactériens dans les infections humaines, les Brucellae et les mycobactéries ne sont pas traitées chez les animaux, mais sont contrôlées par abattage et seuls les campylobacters et les salmonelles représentent un risque normal pour l'homme. Le traitement des premiers est peu valable, de même que celui des entérites dues aux salmonelles qui contaminent les aliments d'origine animale. En Grande-Bretagne, la plupart des Salmonelles multirésistantes aux antibiotiques (généralement S. typhimurium DT 204/204c) restent sensibles au furazolidone et ne posent donc pas de problème au vétérinaire. L'infection, qui atteint surtout les bovins, peut facilement être contrôlée en limitant la circulation des animaux et de nombreuses infections humaines peuvent être évitées par une hygiène culinaire adéquate et la pasteurisation du lait. Il n'y a aucune preuve d'une menace pour l'homme due aux E. coli multi-résistants, car les souches animales n'affectent que très rarement, sinon jamais, l'homme et il n'y a aucune preuve médicale qu'il existe un lien entre les plasmides multirésistants d'E. coli et les souches résistantes chez l'homme.

Les résidus dans le lait et la viande peuvent être réduits, mais non éliminés, par une information adéquate combinée avec des contrôles et des amendes.

ZUSAMMENFASSUNG

ANTIBAKTERIELLE SUBSTANZEN UND RESISTENZ
GEGENÜBER DIESEN SUBSTANZEN BEI DER BEHANDLUNG
UND BEKÄMPFUNG TIERISCHER KRANKHEITEN

David J. Taylor,
Abteilung Veterinärpathologie,
Universität Glasgow Veterinärschule,
Bearsden Road, Bearsden, Glasgow G61 1QH

Zur Behandlung tierischer Krankheiten, zur Verringerung von Schmerzen und Leiden, zum Verhindern der Sterblichkeit und zum Wiederherstellen des Gesundheitszustands und der Produktivität klinisch befallener Tiere innerhalb einer möglichst kurzen Zeit werden weitgehend antibakterielle Substanzen eingesetzt. Eine Reihe nicht verwandter antibakterieller Substanzen stehen in den meisten westeuropäischen Ländern dem Tierarzt für die Behandlung Tierkrankheiten zur Verfügung. Einsetzen kann man diese antibakteriellen Substanzen zum Behandeln von Einzeltieren oder Gruppen von Tieren, die an einer Krankheit leiden, und zwar durch Injektion, durch arzneiliche Behandlung des Trinkwassers oder des Futters in therapeutischen Mengen, oder lokal. Sie können auch zum Verhüten von Krankheiten eingesetzt werden, die sich sonst entwickeln würden, oder zum Eliminieren von Infektionen in einer Gruppe mit Hilfe einer strategischen Behandlung. Einige antibakterielle Substanzen mit wenig Beziehung zu denjenigen, die für therapeutische Zwecke eingesetzt werden, gibt man dem Futtermittel als Leistungsförderer bei.

Die meisten antibakteriellen Behandlungen sind erfolgreich und eine große Zahl umfangreicher Infektionen reagiert noch immer empfindlich auf Penicillin. Die Zahlen, die im Vereinigten Königreich in staatlichen Veterinärlabors bezüglich der Diagnose von Krankheiten ermittelt worden sind, lassen erkennen, daß 18-30 % der in Hornvieh, Schafen und Schweinen diagnostizierten Mikrobeninfektionen von penicillinempfindlichen Organismen verursacht werden und nur 5-7 % sämtlicher Bakterieninfektionen mit unempfindlichen Organismen wie Pseudomonas spp., Mycobacteria und Brucella assoziiert sind.

Die Erfolglosigkeit einer Behandlung kann eine Reihe von Ursachen haben, mit Einbegriff des Nichterkennens von Krankheiten als nichtbakteriell oder des richtigen Erregers, des Fehlens einer rechtzeitigen Behandlung, des Fehlens ausreichender Medikation, einer angemessenen Stütztherapie und der Maßnahmen zum Verhindern einer erneuten Infektion. Wohl tritt eine Resistenz gegen antibakterielle Substanzen auf, aber es gibt nur wenige oder gar keine Syndrome, die sich nicht durch eine Behandlung mit Hilfe des verfügbaren Sortiments antibakterieller Substanzen bekämpfen lassen. Auf lange Sicht kann man zur Bekämpfung tierzuchttechnische, stützende und Impfungsmaßnahmen ergreifen. Einige Beispiele von Syndrome wie Rindermastitis, <u>E. coli</u> enteritis, Erkrankunger der Atmungswege und Schweinedysenterie werden erörtert und die Mittel zum Überwinden der Resistenz gegenüber antibakteriellen Substanzen werden kurz erläutert.

Schließlich bedenkt man das Risiko für den Menschen, das sich aus der antibiotischen Behandlung in Tieren ergibt. Organismen wie <u>Leptospira</u> <u>hardjo</u>, <u>Bacillus</u> <u>anthracis</u>, und <u>Chlamydia</u> <u>psittaci</u> reagieren alle empfindlich auf antibakterielle Substanzen bei Humaninfektionen. <u>Brucellae</u> und Mycobacteria werden in Tieren nicht behandelt sondern durch Schlachten bekämpft und nur Kampylobacter und Salmonellae stellen eine allgemeine Bedrohung für den Menschen dar. Eine Behandlung hat bei den erstgenannten nur wenig Wert wie bei den nahrungsvergiftenden Salmonellae tierischen Ursprungs. Im Vereinigten Königreich reagieren die meisten vermehrungsresistenten Salmonellae (gewöhnlich <u>S. typhimurium</u> DT 204/204c) weiterhin empfindlich auf Furazolidon und sind deshalb für den Tierarzt kein Problem. Die vor allem bei Hornvieh auftretende Infektion könnte leicht durch Bewegungseinschränkungen bekämpft werden und zahlreiche Humaninfektionen könnte man durch eine angemessene Kochhygiene und die Pasteurisierung der Milch verhindern. Es gibt keine Beweise für eine unmittelbare Bedrohung des Menschen durch vermehrungsresistente <u>E. coli</u> da tierische Stämme den Menschen nur selten, wenn überhaupt in Mitleidenschaft ziehen und es in medizinischer Hinsicht keine Beweise für eine Verbindung zwischen vermehrungsresistenten <u>E. coli</u> Plasmiden und menschlichen vermehrungsresistenten Stämmen gibt.

Antibakterielle Rückstände im Fleisch und in der Milch können durch eine Kombination von Erziehung, Überwachung und finanziellen Strafen reduziert aber nicht ausgemerzt werden.

ANTIBACTERIALS AND RESISTANCE IN MAN

Dr. C. TANCREDE,
Service de Microbiologie Médicale,
Institut Gustave Roussy,
Villejuif - France

The intestine of man and animals is the largest reservoir of bacteria
in the environment. It plays a most important role in the contamination of
food from animal origin and in the transmission of bacteria from the animal to
man and vice-versa. The number of bacterial cells per gram of matter is
estimated to be 10^{11} in the end of the intestine. These are mainly strict
anaerobic bacteria which belong to many species. The microbial pattern of the
intestinal flora varies depending on the host. The digestive tract and its
flora constitute an ecosystem where there are multiple interactions : host-
bacteria, food-bacteria, bacteria-bacteria, etc.(Ducluzeau and Raibaud, 1979).
The presence of active antibiotic concentrations in this ecosystem leads to
modifications. The selection of a resistant bacterial strain is not as easy
to study there as _in vitro_, since many additional factors intervene, such as
the interaction amongst micro-organisms. Indeed the development of each
bacterial strain is not uncontrolled as if it were alone in a culture : the
population size of each strain is limited by the whole ecosystem. In this
context, the effect of the antibiotic on a strain X is the result of its
activity on this strain and its activity on all the micro-organisms present in
the ecosystem which also exert their activity on strain X. ˙The future of a
resistant strain and the risk of contamination which it may represent will
depend firstly on the action of the selective antibiotic on the ecosystem and
the action on the strain of the ecosystem which has been modified by the
antibiotic (Tancrède et al., 1977). Secondly, when the antibiotic is no
longer administered, the future of the resistant strain depends on its
adaptation to the ecosystem which opposes a more or less important resistance
to colonisation. So the type of diffusion of a resistant strain depends on
its more or less good adaptation to the ecosystem where it develops (Raibaud
et al, 1977).

Generally speaking, beyond the resistant bacteria, it is the genetic
determinant of resistance which must be taken into consideration : apart from
the mutation performed on the chromosomes of the bacteria which will only lead
to vertical transmission from mother-cells to daughter-cells, the genetic

resistance determinants may be more or less independent from the bacteria and
may be transmitted horizontally from cell to cell. This is the case during
transformation, transduction by bacteriophage and conjugation processes. The
latter phenomenon has a considerable importance in the diffusion of resistance
in gram-negative bacilli which are important in human and veterinary pathology. The transmissible genes may be localised on the chromosome or on a
plasmid. In the latter case, the possibility of interspecies transfers or
transfers between phylogenically different bacteria represents a considerable
potential of resistance spreading. These transfers of genetic determinants
can be observed in vitro. They also occur in vivo, in the intestine of man or
animals. Even if the frequency of these transfers is very low, the consequences may be very important if the new host bacteria is well adapted to the
ecosystem and ensures a large diffusion of the genetic resistance determinant.
Now it has been proved that the simple digestive transit, without implantation, of a giver bacteria suffices to perform a plasmid transfer to a receiver
bacteria which is very well implanted in the intestinal flora (Fig. 1).

Fig. 1 : Transfer of resistant plasmids to heteroxenic mice with human flora.

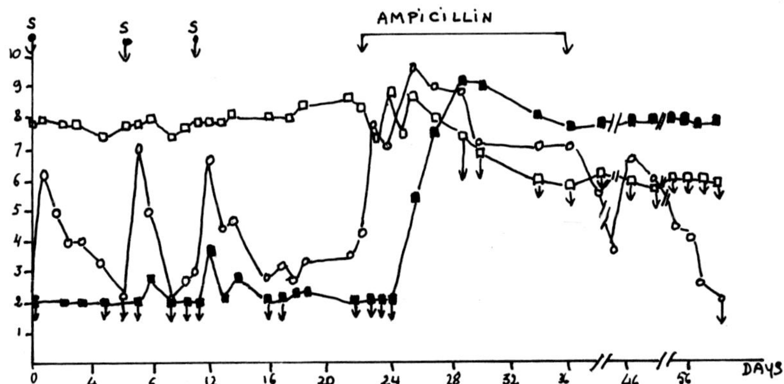

The symbols indicate the detection limits of the corresponding
strains o , □ and ■ .

The giver strain Serratia liquefaciens (o) does not become implanted, in spite
of repeated administrations. The receiver E. coli strain (□) is well
implanted. Transconjugating E. coli (■) appear and become the dominant
enterobacterial strain after treatment with ampicillin. (Duval-Iflah et al,
1980).

Whatever the origin of the genetic resistance determinant may be and
however infrequently found in nature, it is thus possible that once it is
transferred on to a well adapted vector bacteria, it may be found increasingly
frequently and on bacteria responsible for infection in man or animal.

These scientific data ought to be compared to what is observed in practice in man or animal. Since antibiotics have been used, the emergence of resistant strains has been observed in most bacterial strains. The frequency of isolation of these strains is much greater where antibiotics are widely used (Gedek, 1981) : hospital environment for man, intensive farming units for animals. However, in man as well as animal, it has been observed that subjects which had had no antibiotic medication carried resistant bacteria (Fig. 2).

Fig. 2 : Subdominant populations of resistant enterobacteria in 100 normal persons who had received no antibiotic.

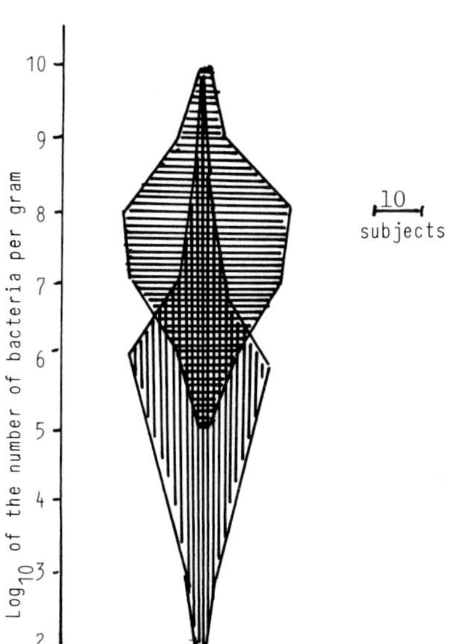

Antibiotics found
Frequency classification

A S K C T SU	31	S T	2
T	29	S SU	2
A C T SU	18	K T	2
A S C T SU	17	A S SU	2
S T SU	14	A K T	2
A	13	A C SU	2
A T	12	K T SU	2
T SU	11	A S K T	2
A T SU	10	A S C T	2
A S T SU	9	A S K C T	2
C T SU	8	A C	1
C T	5	A SU	1
S C T SU	5	K SU	1
A K C T SU	5	C SU	1
A S	4	A S T	1
A C	4	S K T	1
S K C T SU	4	S C SU	1
K	3	K C T	1
K C T SU	3	K C SU	1
A S K C SU	3	A S C SU	1
S	2	A K C T	1
		A K C SU	1

Percent of individuals with enterobacteria resistant to tested antibiotics

Tetracycline	85
Sulphonamides	75
Ampicillin	73
Chloramphenicol	65
Streptomycin	64
Kanamycin	49
Gentamycin	0

93 % of persons carry enterobacteria resistant to at least one of the six antibiotics : ampicillin (A), streptomycin (S), kanamycin (K), chloramphenicol (C), tetracyclines (T), sulfonamides (S) (Lemozy, 1976).

This indicates the wide dissemination of resistance determinants for which one can minimize in man the following influences :
- the role of earlier antibiotic treatments;

- the transmission from man to man of resistant bacteria;

- and also contamination from animal origin. The antibiotic treatments prescribed in human medicine are by far the most important element, but one cannot neglect the role of antibiotics used in animals intended for consumption.

The thought that a limitation in the use of antibiotics could reduce the incidence of resistant strains is legitimate. One could, however, imagine that many genetic resistance determinants would remain present for a very long time, even indefinitely in the vector bacterial strains which are well adapted to the human or animal host (Richmond, 1981). In any case, the immense benefit brought by antibiotics to human health and the production of animals for human consumption is so important that one could not envisage to question it.

There remains the perspective of a more rational utilisation of antibiotics, which would limit the spreading of resistance : one could reasonably propose in human medicine a more controlled use of antibiotics, but it is difficult to define precisely the limits of useless prescriptions.

In the animal field, an important question is to define the active residual concentrations of antibacterial substances which are able to select resistant strains capable of playing a role in man (Finland, 1975, and Lacey, 1981). The _in vitro_ study of the effect of subinhibitory concentrations of an antibiotic on the equilibrium between a sensitive bacterial population and the population of the same bacteria carrying a resistance character may show that concentrations significantly lower than the minimum inhibitory concentration have an unfavourable effect on the sensitive bacteria and favour the development of the resistant strain (Ezrow et al, 1979, and Lebek and Egger, 1983). This does not mean that the same concentrations will have the same effect in the ecosystem of the digestive tract, since the action of the antibiotic on the remainder of the flora and the effect of the flora on the studied bacteria must be taken into account simultaneously.

An experimental model permits us to observe _in vivo_ the role that subinhibitory concentrations may play in the large reservoir of bacteria which is the intestine of a mammal. This consists in maintaining axenic mice in an isolator and implanting in their digestive tract a sample of faecal flora from another host : human, pig, etc... (Fig. 3). In these conditions, the same equilibrium amongst bacterial populations as that observed in the original host is reproduced and it is possible to study simultaneously the interaction between bacterial strains and the action of an antibiotic _in vivo_. The introduction of the selected bacterial strains, whether carrying resistance determinants or not, enables us to observe the resistance of the ecosystem to the colonisation by these exogenous bacteria in the presence or absence of an antibiotic. It is thus possible to study in the ecosystem the action of the antibiotic on the future of the bacteria and the resistance determinants.

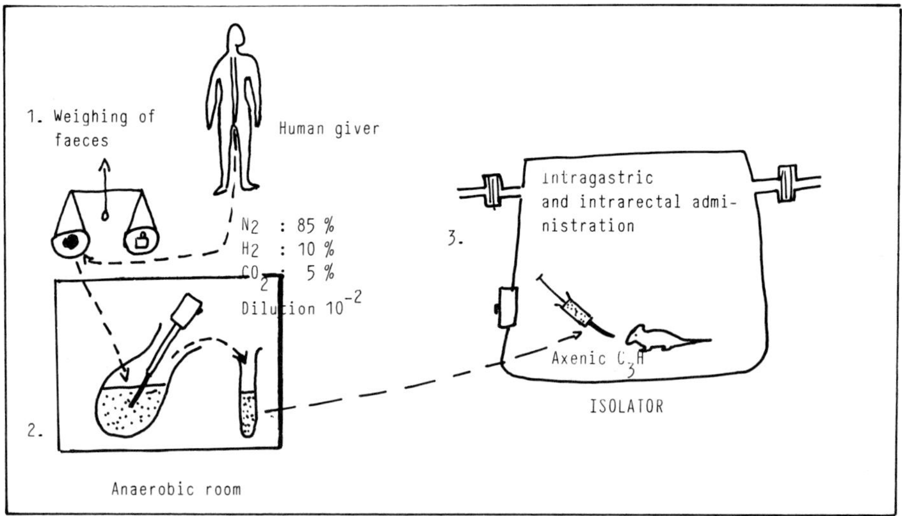

Fig. 3 : Heteroxenic mice with human flora.

The intestinal flora of the giver is reproduced in the mouse :
- quantitatively : the total count of intestinal bacteria is the same in the man and in the mouse,
- qualitatively : the dominant bacterial strains are the same,
- functionally : the barrier effects and resistance to colonisation observed in the giver are reproduced in the intestines of the mice.

This experimental model is proposed for the determination of antibacterial concentrations which may have significant consequences on the selection of resistance factors in man or various animals (Andremont et al, 1983, and Corpet, 1980).

This prospective view of the possible inconveniences related to antibacterial substances should be compared to the observations performed on the control : counts of bacteria carrying resistance factors in human and animal faeces, both with and without antibiotic treatment. The comparison of strains isolated in this manner and their resistance character would form the basis of epidemiologic studies on resistance.

The results of such controls would enable us to define what are the antibacterial products which are most often responsible for the diffusion of resistance factors and what concentrations lead more particularly to this risk.

The in vivo observation of antibiotic effects on the human and animal flora is indispensable for the preparation of rational use recommendations if one wants to achieve the double target of limiting the diffusion of resistant bacterial strains and maintaining and improving the considerable progress due to antibiotics both in the field of health and economy.

182

BIBLIOGRAPHY

Andremont, A., Raibaud, P. and Tancrède C., 1983. Effect of Erythromycin on
    Microbial Antagonisms : A study in Gnotobiotic Mice Associated with a Human
    Fecal Flora. J. Infec. Diseases 148 (3): 579-587.
Corpet, D., 1980. Influence de faibles doses de Chlortétracycline sur la
    résistance à la Chlortétracycline de Escherichia coli dans le tube digestif
    de souris axéniques hébergeant des flores complexes d'enfant, de veau ou de
    porcelet. Ann. Microbiol. (Inst. Pasteur) 131 B: 309-318.
Ducluzeau, R. and Raibaud, P., 1979. Ecologie microbienne du tube digestif.
    Ed. Masson.
Duval-Yflah, Y., Raibaud, P., Tancrède, C. and Rousseau, M., 1980. R-plasmid
    transfer from Serratia liquefaciens to Escherichia coli in vitro and in
    vivo in the digestive tract of gnotobiotic mice associated with human fecal
    flora. Infection and Immunity 28 (3): 981-990.
Ezrow, L., Saceanu, C. and Dabydeen, H., 1979. Range of Antibacterial Acti-
    vity of Antibiotics at Subminimal Inhibitory Concentrations : The Ratio of
    Minimal Inhibitory Concentration to Minimal Antibiotic Concentration.
    Reviews of Infections Diseases 1 (5): 821-824.
Finland, M., 1975. Relationship of antibiotics in animal feeds and salmonel-
    losis in animals and man. J. of Animal Sciences, 40 (6): 1222.
Gedek, B., 1981. Factors influencing multiple resistance in Enteric Bacteria
    in animals. In: Ten Years on from Swann. D.W.Jolly, D.J.S.Miller, D.B.
    Ross, P.D.Simm, Ed., Royal College of Veterinary Surgeons, Wellcome
    Library, London, pp. 111-126.
Lacey, R.W., 1981. Are resistant bacteria from animals and poultry an
    important threat to the treatment of human infections ? In : Ten Years on
    from Swann. D.W.Jolly, D.J.S. Miller, D.B. Ross, P.D. Simm, Ed., Royal
    College of Veterinary Surgeons, Wellcome Library, London, pp. 127-144.
Lebek, G. and Egger, R., 1983. R-Selection of Subbacteriostatic Tetracyclin
    Concentrations. Zbl. Bakt. Hyg. A 255: 340-345.
Lemozy, J., 1976. La résistance des bactéries aux antibiotiques en population
    urbaine. Sc. Tech. Anim. Lab. 3 : 170-171.
Raibaud, P., Ducluzeau, R., Tancrède, C. L'effet de barrière microbien dans
    le tube digestif : Moyen de défense de l'hôte contre les bactéries
    exogènes. Médecine et Maladies Infectieuses 7 : 130-134.
Richmond, M.H., 1981. The emergence of antibiotic resistance in Bacteria and
    its implications for antibiotic use. In : Ten Years on from Swann.
    D.W.Jolly, D.J.S.Miller, D.B.Ross, P.D.Simm, Ed., Royal College of
    Veterinary Surgeons, Wellcome Library, London, pp. 99-109.
Tancrède, C., Azizi, P., Raibaud, P. and Ducluzeau, R., 1977. Conséquence de
    la destruction des barrières écologiques de la flore du tube digestif par
    les antibiotiques. Perturbations des relations entre l'hôte et les bacté-
    ries potentiellement pathogènes. Médecine et Maladies Infectieuses, 7
    (2bis) : pp. 145-149.

RESUME

<u>ANTIBACTERIENS ET RESISTANCE CHEZ L'HOMME</u>

Cyrille TANCREDE,
Service de Micriobiologie Médicale,
Institut Gustave-Roussy,
Villejuif - France

L'intestin de l'homme et des animaux est le plus important réservoir de bactéries de l'environnement. Il joue un rôle capital dans la contamination des denrées d'origine animale et dans la transmission des bactéries de l'animal à l'homme et vice-versa. Le tube digestif et sa flore constituent un écosystème, où de multiples interactions (hôte-bactéries, aliments-bactéries, bactéries entre elles) influencent le développement de chaque souche de bactérie, l'action des antibiotiques et la sélection et le développement de souches résistantes. Le risque de diffusion d'une souche résistante va dépendre de son adaptation à l'écosystème, au sein duquel elle se multiplie.

Depuis que les antibiotiques sont utilisés, l'émergence de souches résistantes dans la plupart des espèces bactériennes a été constatée. La fréquence de l'isolement de ces souches est beaucoup plus grande, là où il est fait un large emploi d'antibiotiques : milieu hospitalier chez l'homme, centre d'élevage intensif pour les animaux. L'idée, qu'une limitation de l'emploi des antibiotiques pourrait amener une raréfaction des souches résistantes, est légitime. On peut penser, toutefois, que nombre de déterminants génétiques de résistance resteraient présents très longtemps, voire indéfiniment, dans des populations bactériennes vectrices bien adaptées à l'hôte humain ou animal. En tout état de cause, le bénéfice immense, qu'ont apporté les antibiotiques à la santé humaine et à l'élevage des animaux destinés à la consommation, est tel qu'il n'est pas pensable de le remettre en question. Reste la perspective d'une utilisation plus rationnelle des antibiotiques qui limiterait la dissémination de la résistance !

Les éléments à prendre en considération sont les déterminants génétiques de résistance. Ils sont plus ou moins indépendants de la bactérie et peuvent se transmettre horizontalement à d'autres cellules. Ces transferts sont observables <u>in vitro</u>, mais ce type d'étude est insuffisant, car les concentrations observées n'y ont pas nécessairement le même effet que dans l'éco système du tube digestif, à cause des diverses interactions qui s'y produisent. Un modèle expérimental, par lequel on reproduit, chez l'hôte expérimental, le même type d'équilibre de chez l'hôte initial, permet d'étudier simultanément les interactions entre souches bactériennes et l'action d'un antibiotique <u>in vivo</u>. Enfin, la comparaison de ces observations avec des souches

isolées sur le terrain (recherche du portage de bactéries pourvues de déterminants de résistances dans les fécès d'humains et d'animaux, traités ou non par
l'antibiotique) constituerait la base d'études épidémiologiques de la résistance.

Les résultats de tels contrôles permettraient de préciser quels sont
les produits antibactériens plus particulièrement responsables de la diffusion
de caractères de résistance et quelles sont les concentrations qui font courir
ce risque. L'observation in vivo des effets des antibiotiques sur la flore de
l'homme et des animaux est indispensable à l'établissement de recommandations
rationnelles d'emploi, si l'on veut atteindre le double but de limiter la diffusion de souches bactériennes résistantes et d'améliorer les progrès considé-
rables permis par les antibiotiques tant dans le domaine de la santé que dans
celui de l'économie.

ZUSAMMENFASSUNG

ANTIBAKTERIELLE SUBSTANZEN UND RESISTENZ IM MENSCHEN

Cyrille Tancrède
Abteilung Medizinische Mikrobiologie
Institut Gustave-Roussy
Villejuif, Frankreich

In der Umwelt fungieren die Eingeweide von Mensch und Tieren als größtes Sammelbecken von Bakterien. Bei der Kontamination von Nahrungsmitteln tierischen Ursprungs und der Übertragung der Bakterien vom Tier zum Menschen und umgekehrt, spielen sie eine höchst wichtige Rolle. Das Verdauungssystem bildet zusammen mit seiner Flora ein Ökosystem, in dem zahlreiche Wechselwirkungen (Wirt-Bakterien, Nahrungsmittel-Bakterien, Bakterien-Bakterien ...) die Entwicklung jedes Bakterienstamms, die Wirkung der Antibiotika sowie die Wahl und Entwicklung resistenter Stämme beeinflussen. Das Risiko hinsichtlich der Ausbreitung eines resistenten Stamms hängt davon ab, wie gut er sich dem Ökosystem anpasst, in dem er sich entwickelt.

Seitdem man Antibiotika verwendet, konnte man das Auftreten resistenter Stämme bei den meisten Bakterienarten beobachten. Die Häufigkeit einer Isolierung dieser Stämme ist viel größer dort, wo man Antibiotika im umfangreichen Maße einsetzt : Krankenhausumgebung bei Menschen und Massenviehhaltung bei Tieren. Der Gedanke, eine Einschränkung des Einsatzes von Antibiotika könne die Inzidenz des Widerstands verringern helfen, ist wohlbegründet. Wahrscheinlich würde die Zahl der Determinanten des genetischen Widerstands in den Trägerbakterien, die dem menschlichen oder tierischen Wirt gut angepasst sind, jedoch auf sehr lange Sicht, wenn nicht sogar auf unbegrenzte Zeit gegenwärtig bleiben. Auf jeden Fall ist der Nutzen der Antibiotika für die Gesundheit des Menschen und die für den menschlichen Verzehr bestimmte tierische Erzeugung so groß, daß man ihn vernünftigerweise nicht in Frage stellen kann. Es bleibt die Aussicht auf einen rationelleren Einsatz der Antibiotika, mit entsprechender Begrenzung des Ausbreitens der Resistenz.

Die Determinanten des genetischen Widerstands sind die Faktoren, die es zu berücksichtigen gilt. Sie sind von den Bakterien mehr oder weniger unabhängig und können horizontal auf andere Zellen übertragen werden. Diese Übertragungen lassen sich in vitro beobachten, aber diese Art der Untersuchung reicht nicht aus weil die beobachteten Konzentrationen nicht notwendigerweise die gleiche Wirkung wie im Ökosystem des Verdauungstraktes haben, und zwar wegen der verschiedenen Wechselwirkungen, die darin stattfinden. Ein Experimentalmodell, bei dem die gleiche Art von Ausgleich wie im ursprünglichen Wirt dann im Experimentalwirt erzeugt wird, ermöglicht die gleichzeitige Untersuchung der Wechselwirkungen zwischen Bakterienstämmem und der

Wirkung des Antibiotikums _in vivo_. Schließlich würde der Vergleich dieser
Beobachtungen mit den im Feld isolierten Stämmen (Zählung der resistenz-
tragenden Stämme, die in menschlichen und tierischen Fäkalien festgestellt
werden, ob sie behandelt worden sind oder nicht) die Grundlage epidemiolo-
gischer Resistenzuntersuchungen bilden.

Mit Hilfe der Ergebnisse solcher Kontrolle könnte man feststellen,
welche antibakteriellen Substanzen im besonderen für die Ausbreitung der
Resistenz verantwortlich zeichnen, und in welchen genauen Konzentrationen.
Die Beobachtung _in vivo_ der Auswirkungen von Antibiotika auf die menschliche
und tierische Flora ist im Hinblick auf das Ausarbeiten von Empfehlungen be-
züglichen ihres rationellen Einsatzes notwendig, wenn man das doppelte Ziel
einer Einschränkung des Ausbreitens resistenter Bakterienstämme und einer
Verbesserung der bedeutenden Fortschritte erreichen möchte, die von den Antibiotika in gesundheitlicher und in wirtschaftlicher Hinsicht ermöglicht
worden sind.

SAFETY ASPECTS IN THE DEVELOPMENT OF NEW
ANIMAL HEALTH PRODUCTS AND THEIR CONSEQUENCES

P. Altreuther,
Pharmaceutical Research Center,
Bayer A.G.,
D-5600 Wuppertal 1, Germany.

SAFETY CONSIDERATIONS IN INDUSTRY HAVE SEVERAL ASPECTS :

first,   those of science, as presented by the experts, requiring adequate scientific potential and a critical mass to survive

second, those of economy, because science is costly and there has to be at least a chance that the money spent on research can be earned in a worldwide market - "Honny soit qui mal y pense"

third,   those of emotion, creating political movements which in turn force governments to strengthen their legislative efforts into more and more restrictive drug laws

and,   last but not least, those of responsibility to the public, it be the user or the consumer.

I shall not discuss the scientific issues again as this has been done by the experts and I shall not stress the field of responsibility. Both are preconditions of work for the industry actively engaged in research. The topic of this presentation will therefore be the economic and political consequences of safety.

The Food and Drug Administration in the U.S.A. is certainly not notorious for ignoring safety aspects in the registration of new drugs. It is much more an institution whose strict guidelines have strongly influenced the development of pharmaceutical legislation in many countries, and whose safety standards are still regarded by many countries as the goal towards which their own legislation should strive. What, then, could it mean when the Director of the Bureau of Veterinary Medicine, Dr. Crawford, is now saying that "it will be a long time, if ever, before we have all the drugs we need" and when he is justifying by this statement the fact that new ways must be found to speed up the resgistration process ? Certainly this does not imply that suddenly an uncritical approach is advocated in place of the safety guidelines. However, it may well mean that, instead of the occasionally voiced claim that we have too many drugs, we have perhaps too few. Not only do problems remain to be solved, but new and better ways of approaching old problems are demanded. This is why the industry must not be allowed to cut down on research, and the society's need to be protected must not be allowed to turn registration into an unattainable goal. This means that, with its endeavours to develop new active substances in the sphere of veterinary drugs and agents increasing the efficiency, the industry finds itself in an area of conflict, where emotions often threaten to overwhelm the scientific aspects. The two sides must be fairly balanced if the whole is not to suffer.

What is it that has given rise to the situation in the U.S.A. that Crawford criticizes and has started energetically to combat, and that we are now attempting to bring about in Europe ?

We cannot go here into the details of the situation in which lawyers representing consumers successfully take large institutions to court for damages running into millions of dollars or make the state liable for injuries or harm that it could not have prevented by law or decree. However, it has contributed to the fact that the Delaney clause -- the fundamentally understandable proscription of carcinogen containing foodstuffs -- has turned into a registration ruling known as the "Sensitivity of Methods".

Known as the S.O.M. and derisively referred to as "Seldom Operational Methodology", during the period of its experimental application it prevented the registration of virtually <u>all</u> active substances falling with its terms of reference. A substance could be banned on the basis of purely external, structural criteria, such as the presence of nitrogen in an aromatic ring or in a heterocycle, or similarity to an active substance known to be carcinogenic. The investigations that consequently became necessary were, however, so difficult and so lengthy that a definitive refutation of the suspicion or satisfaction of the safety conditions was extremely unlikely.

A decree that makes demands going beyond the current state of scientific knowledge, or requires the application of methods for which neither experience nor suitable models are available, cannot but hinder the registration of a drug. It would make the approval of a substance a game with incalculable risks that industrial research alone would have to bear. The consequence would be the abandonment of research in the face of the risks, and a number of companies have indeed come to this obvious conclusion in recent years and have opted out of their "animal health" activities.

It is not only the U.S.A. that has to suffer from the consequences of its earlier regulations -- the States have begun to solve their problems. One problem that still persists in Europe is the attitude towards the use of antibiotics in animals, and not only for promoting growth but even for therapeutic purposes. The Swann Report of 1969 is the scientific basis for this attitude, which has subsequently led to a ban on the addition of penicillins and tetracyclines to animal feeds. I am not objecting to this step ; the current state - by that time - of scientific knowledge makes it at least understandable, although other countries, such as the U.S.A., do not believe that these measures have to be taken for the protection of their citizens. But where do we stand today ?

In mid-December of last year the American Council on Science and Health (ACSH) published a report -- following a preliminary report by the U.S. National Academy of Sciences in 1980 and a symposium "Ten Years on from Swann" that took place in London in 1981 -- which illuminated the dangers of the subtherapeutic use of antibacterial substances in animal feeds.

<u>Conclusion</u> : Despite 30 years of use of these active substances and 15 years of research aimed at identifying the real dangers, no damage is actually evident. Allow me to cite one Dr. Foster, a member of the ACSH and the Director of the Food Research Institute of Wisconsin University :

"The dire consequences that have been predicted for the past 15 or 20 years simply have not come about. As with so many other safety questions, we are left wondering whether to act on a theoretical possibility of danger or continue a valuable practice that has shown no harm in three decades".

The question posed applies to the U.S.A. ; in Europe we have already been acting for years on the basis of a theoretical possibility of danger, and this cannot be faulted. However, how likely must a danger be to force us to accept serious disadvantages ? Can we expect the citizen to tolerate every new burden because of a dubious enhancement of his safety ? In Europe we have never worked out the economic consequences of the proscription of penicillins and tetracyclines. The ACSH study puts the yearly loss at around 3.5 thousand million U.S. dollars -- a figure that is hardly likely to be any smaller in Europe.

Is it right that in Europe scientific reason should play no part at all when -- alledgedly -- increased safety is involved, and are we not duty-bound to prove that safety is truly increased ?

And what are the consequences on the part of the legislators when a supposed risk turns out to be an illusion ? Do laws have to become more and more restrictive, or is the opposite also possible when new knowledge permits ? These are questions to which answers must be found by discussions between all interested parties. On the topic of anti-bacterial substances in animals, for example, Prof. Richmond, who with his research helped to lay the foundations for legal restrictions of their use, said at the "Ten Years on from Swann" symposium :

"... one must have the courage to use the agents to the limits of their potential. Because of general unease over the future fate of antibiotics and an understandable preoccupation with human health, a strongly protectionist attitude has been allowed to develop ; and this has tended to bring the "regulators" and the "merchandisers" into sharp conflict. There seems little doubt that by cooperation rather than by confrontation a series of measures could be adopted which would allow a more extensive use of antibiotics for all purposes without seriously jeopardising their long term efficacy. This would be of much value to all concerned. It does not seem too much to hope for that a rational approach to the problem could be developed".

It is to be hoped that Richmond will be proved right if development is to remain possible. The capital investment put up by the individual companies on development alone is, after all, considerable : in 1983 the Animal Health Institute estimated the costs of developing an active substance intended for use as an efficiency promotor in cattle at 15 million dollars ; it was also said that this development took about 8 years to complete. A fermentation product for ruminants does not come any cheaper in Europe.

Naturally enough development costs vary widely, depending on the type of the product. However, what is certain is that no product intended for use in animals for the food industry costs less than 5 million DM to develop, while in the upward direction the sky is the limit. The average cost may be between 8 and 15 million DM for a synthetic product which it can take anything from 4 to 8 years to develop. Now add the time required for the registration procedure. The expenditure on actual clinical trials averages some 10 % of the total costs ; between 40 and 70 % of the investment is spent on safety tests. Thus, a sensible development of such safety-based legislation in the various countries becomes all the more important.

In the EEC with the 81/851 and 81/852 guidelines a common base has been created for the development and registration of veterinary drugs. This must be welcomed without any reservations and its full incorporation into national legislation is to be hoped for. Opportunities for individual assessment, which inevitably arise in a settlement between several partners, will hopefully not result again in the registration procedure becoming more difficult and more prolonged.

Experience from the sector of human medicine, however, suggests that it is higly unlikely that approval will be granted in at least 5 member countries whithin the 540 days possible. If the specifications have indeed truly been brought into line, why cannot registration in one member country lead automatically to registration in all the other countries ? That and only that would mean true harmonization and partnership.

I shall refrain from going in greater detail into the individual points of these guidelines, which are based more on nationalism than on a scientific basis ; as e.g. into the test for residue toxicity after oral administration which had to last 3 to 6 months, or the requirement for studies into allergic manifestations, for which, according to the present state of scientific knowledge, there are as yet no obligatory methods.

Instead, I should like to turn my attention to some questions connected with the regulations proposed in these guidelines for drugs administered with the animal feeds, which in due course -- if they should be passed -- will make it considerably more difficult to develop veterinary drugs and will hinder their registration. I am talking about the list of active substances that may be used in premixes for drugs to be administered with the feed. Such active substances, according to one of the member countries, should undergo a further selection procedure and a special list should be compiled. If the whole concept of one formulation being more strictly controlled than all the others is unusual -- no such restrictions exist in the U.S.A. -- then it appears completely unnecessary that a formulation, after its active substance has been reviewed successfully by the relevant authorities, should have to go through a second registration and selection procedure. This second procedure -- a selection process -- would inevitably be based on irrelevant, economic, political, or merely personal notions (for scientific criteria have already been utilized in the course of the active substance's registration in other countries) -- a situation which would be quite unacceptable for the registration of a drug.

Happily, as was definitively stated in the Report of the Committee to the Council Com (83) 367, most of the experts consulted by the Committee are of the opinion that the existing guidelines are adequate. Despite this, the discussions are still continuing.

The principal argument in favour of the "positive list" is the idea that mass treatment cannot be controlled correctly unless the number of active substances approved in the Community is strictly limited. However, mass treatment is not only implemented via animal feeds but is also possible by the oral route by individual administration in the form of a drench, by administration in the drinking water, or by the parenteral route as a pour-on formulation. All these are perfectly feasible. Problems other than those of adherence to withdrawal times, which arise after treatment with all veterinary drugs, do not emerge as a result. One might even consider it simpler to monitor a whole herd rather than individual animals in this herd.

Moreover, the selective list is conceived of as a list of "generics". As is known from experience with the list of feed additives, this gives manufacturers of chemicals an opportunity to bring active substances as chemicals directly onto the -- grey ? -- market, with total disregard for pharmaceutical regulations and without the specialized knowledge required in the drug laws. The net result is that not only does the research industry, which has invested money in the product and considerable scientific effort at advisory level and for further development of the technology, finds itself devoured by competitors, but risks arise for the authorities and the consumers alike : does the product conform to the purity criteria, the bio-availability, and the quality of the originally approved material ? It does not seem particularly sensible to want to repeat a mistake that after years of discussion in the animal feed additives sector we were about to eliminate -- even if this were to be done with a presumably unsuitable method. An analytical monograph of additives is only a substitute for the registration of specialities, which alone is acceptable because of the assurances it affords. Here too national egotism has stood in the way of reaching the optimal solution to the problem.

The fact that a positive list would from the outset limit the therapeutic possibilities and handicap research in this sector is a consequence that is also stressed in the Report of the Committee to the Council. It is difficult to see why this matter is still being discussed.

There are safety aspects that make legislation more and more restrictive. As was shown in the example of the S.O.M. and the antibiotics regulations, restrictive legislation inhibits research -- one cannot do research in areas with virtually no access to the market, and as a result society suffers. However, it is society that makes the politicians formulate legal restrictions for its sake. How can we break through this vicious circle, whose adverse effects will be more palpable in Europe than they were in the U.S.A. because of Europe's many small markets and lack of harmonization, even within the EEC ? How can we prevent the loss of competitiveness in a sector that is vital for the existence of countries with poor mineral resources ? Whereas it would be presumptuous to claim that there were ready-made answers to all the questions, a few points are worth considering :

Proper information of the public could take the emotional charge out of the safety discussion, and it is this that gives rise to overreactions and thence to overregulation : consider the example of the ban on the use of anabolic substances in the animal-fattening industry. This came about as a result of the detection of DES in food products, even though the DES had not got into the food as a result of the treatment lege artis but because of a breach of the law. However, it does not make things any harder for the lawbreakers if the regulations applicable to the registration or use of a substance that is banned anyway are tightened up by a whole group -- namely hormones and their analogues -- being included in the ban without due consideration of the pros and cons. What would make things more difficult would be stricter control of the forbidden active substances in all the countries of the Common Market.

It would also help to objectivize the discussion if the analytical detection of a substance was not equated with the harmful effect that it can exert in toxic doses. We reel off ppm's, ppb's, and ppt's as effortlessly as pounds and kilogrammes. Detection is detection, whether it be in picogrammes, microgrammes, or milligrammes : everything is equally dangerous. And then occasionally with half-truths, such as animal experiments which are not extrapolable to man, reference is made to the incalculability of a risk. Explanations are necessary, but industry is not in a good position to do this, since as a representative of public interest it has less credibility than public institutions. Centres of higher education and in particular agencies set up as guardians of public health would be suitable bodies to help in this task.

That industry is interested in instructing its customers in the proper use of active substances in animals and in observing the withdrawal periods is evident from a campaign by the Animal Health Institute, where meat and dairy producers are advised of the devastating consequences of illegal or incorrect use of veterinary drugs : "What would happen if certain key animal health products had to be withdrawn from the market because of improper use :

"half of all broiler chickens would die ; total meat production would decline 16 % ; feed consumption per pound of weight gain would increase 4 % in beef, 7 % in swine and turkeys and 13 % in broilers".

Industry is quite prepared to enter into collaboration and discussions with all who have to bear the risks, costs, or public responsibility.

Take advantage of this offer.

RESUME

SECURITE ET DEVELOPPEMENT DE NOUVEAUX PRODUITS

Peter Altreuther,
Wuppertal-Elberfeld.

Il n'y a pas trop de médicaments vétérinaires et pour citer le Docteur Crawford de la F.D.A. (Food and Drug Administration, Etats-Unis) : "Il faudra encore longtemps avant que nous n'ayons tous les médicaments dont nous avons besoin".

L'industrie pharmaceutique se trouve, en ce qui concerne le développement de nouveaux produits vétérinaires, au centre de tensions émotionnelles, politiques et scientifiques. Des règlementations continuellement plus sévères tant en Europe, suite à la publication du rapport du Comité Swann, qu'aux Etats-Unis freinent le développement de tels produits par l'accroissement des investissements nécessaires (8 à 15 millions de DM) et l'allongement du temps de développement (4 à 8 ans). Dans ce contexte, les directives 81/851 et 852 sont discutées ainsi que la proposition d'une liste positive de produits autorisés dans les aliments médicamenteux.

Il est en outre indispensable que l'industrie assume la responsabilité d'une information complète sur l'usage correct de ses produits.

ZUSAMMENFASSUNG

SICHERHEITSASPEKTE IN DER ENTWICKLUNG
NEUER PRODUKTE UND IHRE KONSEQUENZEN

Peter Altreuther

Wuppertal-Elberfeld

Tierarzneimittel sind nicht überflüssig : "It will be a long time, before we have all the drugs we need" (Crawford, FDA).

Mit der Entwicklung entsprechender Wirkstoffe aber befindet sich die Industrie in einem Spannungsfeld mit emotionalen, wissenschaftlichen und politischen Komponenten. Anhand von Entwicklung in USA (Modifizierung der SOM-Regelung) und in Europa (Konsequenzen des Swann-Reports) werden die Konsequenzen einer nur restriktiven Gesetzgebung aufgezeigt, welche die erhebliche Investition der Industrie für eine Entwicklung (8 - 15 Mio DM für ein synthetisches Produkt in 4 - 8 Jahren Entwicklungszeit) nicht rechtfertigen. In diesem Zusammenhang werden die Richtlinie 81/851 und 852 kommentiert und der Vorschlag der Positivliste für Wirkstoffe in Fütterungsarzneimitteln diskutiert. Die Verantwortung auch der Industrie für sachgerechte Information und besonders auch für die sachgerechte Anwendung ihrer Produkte wird hervorgehoben.

DISCUSSION

Question : What is the normal microflora of the duodenum, jejunum and ileum ?

C. Tancrède : In the upper part of the digestive tract two phenomena limit the bacterial population. These are chemical and mechanical factors. Firstly the normal gastric acidity considerably reduces by a factor of one to ten thousand the number of bacteria which enter the digestive tract. The mechanical factor is the speed of transit in the upper small intestine. This is why, in the duodenum in particular, the number of bacteria is very small. It increases slightly as one moves down to the coecum. The bacteria that are found there are usually facultative anaerobes, but there are no resident bacteria in that part of the intestine, with one exception : at microscopic examination one can observe by screening some bacteria which cannot be cultivated. They can be transmitted to a germ-free animal, but one does not know what they are. They have their noses dug in the mucosa, they are present but that is all we know. Numerous experiments have been performed to create an important bacterial development in the small intestine with the appearance of strict anaerobic propulations which become resident. This is the case when making a blind loop. Very quickly a flora comparable to that of the coecum develops. This also occurs in cases of diseases when for mechanical or any other reason the transit in the small intestine has slowed down or stopped. Then there is a population.

Question : Dr. Taylor, what is your opinion on the advantages and inconveniences of the use of veterinary drugs combining several antibiotics ?

D. Taylor : In other words, multiple antibiotics in treatment. I prefer to use single antimicrobials appropriate to the disease present, but in some cases such as trimethoprim sulfonamide combinations there is a reason to use the mixture. When two diseases commonly occur together, for example swine dysentery and Bordetella bronchiseptica bronchopneumonia, then I would consider stategic treatment with two antimicrobials as a means of controlling both and in conjunction with all in - all out husbandry and disinfection, thus reducing the length of treatment to the shortest possible time.

Question : If drugs are teratogenic after one dose, is it true that prolonged administration of a teratogenic drug can result in a variety of effects ?

P. Delatour : Yes, as you can see from the tables, the limits of sensitivity are well known. Within these limits the various sensitive embryonic stages are known precisely - sometimes at one day - so that the so-called teratological calendar has been established for rat and man. This means that if a teratogenic product is administered on day D of the sensitive period, one type of malformation is obtained, and 48 hours later other types of malformations will be obtained instead, which will be different from the first trial. Consequently, it is possible to have different effects with one product at one particular dose and these effects will depend on the timing of administration.

Question : Is there any relationship between structure of benzimidazole compounds and embryotoxicity ?

P. Delatour : The antiparasitic activity and the teratogenic activity are due to completely independent effects. The antiparasitic effect is the result of an inhibition of fumarate reductase while teratogenicity results from the inhibition of the polymerization of the microtubules of the mitotic spindle. Consequently there may a priori be substances which are active on fumarate reductase, but do not inhibit the polymerization of the mitotic spindle during cell division, hence the mitotic effect. From the point of view of the structure that does or does not respond to the teratogenic effect, there are differences. A famous case is that of fenbendazole which is not teratogenic even at high doses, while other compounds are teratogenic at much lower doses. The basis of the explanation which permits us to report the differences is firstly linked to the pharmacokinetic differences of intestinal absorption of the product and secondly the metabolisation of the original product which will not be embryotoxic if it metabolizes rapidly. If the product is metabolized slowly the active product will diffuse on to the microtubules of the mitotic spindle inside the tissues, hence toxic symptoms.

Question : Dr. Roe, you gave the impression in your talk that there was no information on the no-toxic effect level for zeranol. Why did you not mention the 90 days' studies in rats and monkeys ?

F. Roe : In the 90 days' rat study there was a no-effect level for orally administered zeranol which was 0.025 mg/kg/day. At that level there were no effects at all and in monkeys using the end point of vaginal cornification which is argually the most sensitive method for detecting oestrogen-

Left to right : Dr. Cyrille Tancrède, Inst. Gustave Roussy; Dr. David Taylor,
Univeristy of Glasgow, Veterinary School; Prof. Arpad Somogyi, Director of
Veterinary Medicines Department, BGA, Berlin (panel discussion leader);
Prof. Philip Shubik (moderator);

Left to right : Prof. Philip Shubik; Mrs. Burgat-Sacaze, Ecole Nationale
Vétérinaire de Toulouse; Prof. Michiel Debackere, State University of Ghent;
Dr. Francis Roe, Independent Consultant; Prof. Paul Delatour, Ecole Nationale
Vétérinaire de Lyon; Prof. André Rico, Ecole Nationale Vétérinaire de Toulouse.

like effects, there was a no-effect level of 0.05 mg/kg/day in a 90 days'
monkey study.

Question : Dr. Taylor, why was there so much noise around the famous
Swann report, when some years later it had to be recognized that it had been
useless, although it was written in the Veterinary Record that "it was a
success even if it was wrong".

D. Taylor : Swann recently considered that use in therapy had generated resistance and that controls to reduce this would hamper human and veterinary medicine. In my opinion, his original recommendations were made in the
context of a primitive diagnostic and therapeutic situation and at the time it
was probably correct to separate therapeutic from growth permitter usage. But
now with the developed veterinary service and a number of antibiotics, the
problems he highlighted are less relevant.

Question : It is usually said that one must use massive doses of
antibiotics in therapy to avoid the development of resistance. How can it
then be explained that the low doses of antibiotics used as growth promoters
have no effect on the development of resistance and can, on the contrary, even
eliminate resistance plasmid ?

D. Taylor : In my opinion many of the antimicrobials used for growth
promotion are different from those in therapy, so would have no effect and
small doses of any antimicrobial may eliminate plasmids by reducing the viabi-
lity of the bacteria and preventing transfer, expression or multiplication of
the plasmid DNA.

Question : The doses required for antimicrobially active substances
vary distinctly between countries : high dose and low dose countries. How
would this affect the development of resistant strains and what kind of dose
regimes were you referring to ?

D. Taylor : In all cases I was referring to high dose treatments.
Secondly in my opinion low dose treatments are not very satisfactory in
therapy and should be unnecessary in appropriate management systems or be used
for short periods only. Levels of high antimicrobial use occur in low level
countries so that resistant strains arise in both types of countries and very
carefully controlled studies would have to be carried out to say exactly what
the effects of high and low dose management were on the development of
resistance.

Question : How can one prove beyond reasonable doubt that an anthelmintic in practical use presents no teratogenic hazards for man ?

P. Delatour : The best way in which to answer this question is to take an example : that of mebendazole is sufficiently explanatory in this respect. If one analyses residues in the liver of sheep slaughtered for example ten days after treatment with therapeutic doses of mebendazole, some 3.1 ppm of total radioactivity are found. Inside these residues, if the extractible residues are split and one considers only the fraction that because of its chemical nature is potentially toxic, one can see that the daily intake of the man who eats 100 g of liver on one day is 216 thousand times inferior to the acceptable daily dose based on the no effect dose of the most toxic metabolite and the coefficient of 1,000. If that man wanted to eat a quantity of liver that would in one meal give him the equivalent of the no effect dose calculated in the rat, he would have to eat 21 tons of liver in one meal. What could be called reasonable limits is based on the extrapolation of experimental results to man, the experimental definition of a no effect level and the intervention of the coefficient of 1,000. From this point, one can make stupid calculations that will bring one to one tonne. It is ridiculous, but shows nevertheless that there is what can be qualified as a reasonable safety margin.

Question : Is there no risk that the toxicity of a drug is influenced by a non toxic product present in certain diets ? If this is true, there may be in certain diets non toxic products enhancing the toxicity of the drug synergism. There may thus be a risk that the toxicity margin for a drug may be lower than expected on basis of toxicological experiments performed on diet not containing the non toxic product.

A. Rico : I think indeed that one can always imagine that something can increase the toxicity of a substance which is being studied, but I would say that the contrary is also true. It is possible - and has been seen - that some products reduce this toxicity. On the contrary, I would like to say that when one observes the manner in which acceptable daily intakes are calculated one can see that the conditions and basic assumptions are very drastic. The animal receives very high doses every single day without exception for the whole duration of its life. When the results are obtained, we take the most sensitive animal species and consider that man is at least 100 times if not more sensitive than the animal. Here again, it is not true and there are results where the human species was more resistant than the animal species. Consequently, I believe, effectively - and this was my conclusion - that arguments demonstrating that there is a risk can always be found, but perfect safety is a chimera. I think that doing proper work based on experimental toxicological data relating to long animal life-spans will reasonably protect the consumer's safety.

Question : Please define the similarities between testosterone, trenbolone acetate and β-trenbolone.

A. Rico : Considering trenbolone acetate and β-trenbolone, there is no problem. β-trenbolone is the active substance, trenbolone acetate is the substance that is administered. It is the same for testosterone and testosterone acetate, the result is the same. You have an hydrolysis which frees β-trenbolone and it is β-trenbolone which undergoes all the biotransformations. If one now compares testosterone and β-trenbolone, it is obvious that these compounds are different. But there are a number of similarities : both are steroids. One is a natural steroid, the other is indeed a synthetic steroid. Going deeper in the similarities, there is an oxhydryl in β-position. If one observes what occurs exactly in the biotransformations, one can see that the biotransformation of testosterone in cattle gives β-testosterone and that the biotransformation of β-trenbolone gives α - β -trenbolone. If on the other hand one compares the similarities of affinities for androgenic receptors, one can see that testosterone and betatrenbolone have affinities for the same type of receptors. Thus they are not identical compounds, but they have similarities which are far from being negligible. Trenbolone is nevertheless a xenobiotic and this is why a whole series of work has been conducted and the results of which are now available.

Question : When do you have enough information to conclude that a substance is a genotoxic mutagen ?

F. Roe : For some substances where you know their chemical structure and you know that they are electrophilic just on a chemical basis, or if you know that they can or will be metabolized to electrophiles you really do not need any mutagenicity tests at all. You can be really almost 100 % certain that these substances are potentially genotoxic. The problem is if you do not have that information from the structure and you just get a single positive result in an Ames test or an in vitro test for cytogenic potential, then there are lots of reasons why you can get a misleadingly positive result in these single tests. And as you go up the scale you get more confidence as you go from bacteria to mammalian cells and then again from mammalian cells to the in vivo situation. If you really have got a positive in vivo test for clastogenicity then I would certainly be worried and if I got a positive result in an in vivo test for gene mutation I would be very worried. It is not a question that can be answered really in a generic way.

SESSION III: WHAT ARE THE OBJECTIVES OF THE LEGISLATIONS?

CONTRIBUTORS

Prof. Bernt Hoffmann, Ambulatorische und Geburtshilfliche Veterinärklinik, Justus-Liebig-Universität Giessen.

Dr. Gerhard Behm, Secretary General of the FEFANA, Bonn.

Mr. David Rees, Eli Lilly Research Centre, U.K.

Mr. William Montagu, President of Crown Chemical Company Ltd, U.K.

ANIMAL DRUGS - OBJECTIVES OF LEGISLATION

Bernt HOFFMANN,

Ambulatorische und Geburtshilfliche Veterinärklinik,

Justus-Liebig-Universität Giessen,

6300 Giessen.

Legislation is a constitutional process by which a law is passed. A Law or Act can thus be defined as rules of action imposed by an authority and which Society has to comply with. Adequate measures have to be taken to ensure its implementation. In order to better understand the needs for legislation in the area of animal drugs, some fundamental aspects should be considered.

Basic aspects and interrelationships

The first - and rather banal - question to be answered is : Why are compounds exhibiting pharmacological activities used in or on animals ? The four main general indications for this are as follows :
 a) to treat diseased animals;
 b) to prevent disease;
 c) to prevent other undesired effects;
 d) to increase efficiency of production.

Bearing these indications and the total animal population in mind, it can easily be estimated that the demand for veterinary drugs must simply be huge. In a free-profit oriented economy, the indications "to prevent disease" and "to increase efficiency" of food producing animals are particularly important.

The second question to be discussed is then : What segment of the population is basically involved in or affected by the use of animal drugs ? The answer is simple, they are as follows :
 a) the producer of drugs (pharmaceutical industry)
 b) the user, i.e. the farmer
 c) the consumer of food of animal origin.

The number of individual people involved or affected increases almost exponentially from (a) pharmaceutical industry to (c) consumer. In the consumer group, close to 100 % of the population are unavoidably affected by the use of animal drugs. This fact will be an important point in further discussion.

It has to be conceded that each group or category of persons involved has a certain legitimate interest. This interest expresses itself in the following important points (Table 1).

TABLE 1

Veterinary drugs and legitimate group-specific interests.

Group	Specific interest
Pharmaceutical industry	Profitably to produce veterinary drugs, to meet challenges of the market (new developments)
User (farmer)	To dispose of good quality and efficacious drugs, safe for the user and the animals. Condition of use corresponding to practical needs within given economic frame (cost).
Consumer	To dispose of safe and good quality food of animal origin at a reasonable price.

There is thus a complex system of interrelationships between the groups involved and distinct ways of communication or "feedback mechanisms" will develop. When left to themselves in such a system, the regulatory factors which are obvious or clearly recognized will become dominant. These are quality, efficacy, cost, profit, market and price. The system of feedback mechanisms developed would show very little if any direct interrelationships between the pharmaceutical industry and the consumer. In such an "uncontrolled" system, "a priori" hidden factors such as the safety of food of animal origin will be of secondary nature if not neglected.

However, it has to be the basic aim of legislation to guarantee that the legitimate interests or rights of every segment of the population are equally considered according to their importance. In respect to the use of veterinary drugs, it has to be established that the factor "safety of food of animal origin" is adequately considered, leading to the establishment of the necessary feedback mechanisms between the consumer, pharmaceutical industry and user (farmer).

Furthermore, legislation has to define and specify its requirements. The control mechanisms for implementation of the regulations will also have to be laid down in the legislation.

Awareness, pressure and acceptance
____

In democratic conditions, legislation is a multi-factorial process. Within this process it has to be considered as legitimate if groups of interest try to increase the weight of certain factors in their favour. However, this system will only function properly if the rules are strictly observed. Hence it is important that all groups involved or affected have their points raised.

For quite some time the use of drugs in food animals and the question of the significance of possible residues in their meat or products was - apart from a few exceptions - not a big issue to the consumer. Based on the detection limit of the analytical methods available, residues or traces of most drugs applied could not be detected.

Furthermore, the concept of a "zero tolerance" for certain important compounds which was developed and propagated by various regulatory agencies seemed to guarantee an utmost degree of safety.

It has to be recognized that these circumstances and expressions easily lead to a false interpretation of the true situation, particularly by the average consumer, because it may not easily be understood that "zero tolerance" did not mean "no residues" but simply "no detectable residues".

With increasing sensitivity of the analytical methods available more and more "residues" were detected. Rather suddenly it had to be realized that so far food of animal origin deemed to be free of residues turned into food containing residues, a situation the consumer did neither expect nor accept. Consequently, residue aspects as well as public health aspects not only became an important and appropriately discussed issue, but perhaps the dominating one.

Did this not break the equilibrium that the system had found ? In my opinion, not yet, and it will not do so if the process of forming an opinion continues and includes other issues.

General and specific legislation has to consider the legitimate interests and demands of the groups of people involved or affected as was stated above. In respect to public health the demand for safe food is of utmost importance. However, ultimate safety in general cannot be achieved : safety is rather relative and hence a matter of definition.

Safe food should be defined as food which, based on standard consumption habits - will not affect human health. Thus this food may contain residues which have to be considered as "harmless to human health", but may not contain residues which are "not harmless to human health". According to a definition by court, the indefinite term of justice "not harmless to human health" is distinctly separate from the sequence of terms "toxic, dangerous, not undangerous and questionable" in the foreground to prevent risks to human health. If this view becomes and remains one of the fundamentals for decisions, the whole system of drug legislation or drug approval will stay in balance.

There is no question that the strengthened position of the regulatory factor "safe food" within the system of interrelationships and decision making processes will lead to some major and minor consequences affecting the user (farmer) and pharmaceutical industry. These consequences, however, will have to be accepted. The user (farmer) and pharmaceutical industry have to be aware of this.

Outlook on legislation

The objectives of legislation are clear. This leaves us with the last question to be answered : What has been achieved up to date, what system are we exposed to ?

First it has to be regarded as great progress that Council Directive 81/851/EEC on the approximation of the laws of the Member States relating to veterinary medicinal products and Council Directive 81/852/EEC on the approximation of the laws of the Member States relating to analytical, pharmacotoxicological and clinical standards and protocols in respect to testing of veterinary medicinal products have been passed. Thus a basis for common decision is available within the EEC, though it has to be expected that several years will elapse before an adequate degree of homogeneity is reached for practical purposes.

The latter Directive requests presentation of adequate documentation to the responsible national institutions. A product can only be approved if this documentation indicates satisfactory results in relation to the basic requirements of quality, efficacy, safety (tolerance) for the animal, safety for public health, including implementation of the conditions of use laid down.

Each of these points raises numerous other points; their discussion would be way beyond the scope of this discourse. Nevertheless a few words should be said on safety or public health aspects.

The enigma of safety

Safety is a vague concept, it is a matter of individual interpretation. To achieve an optimum level of safety - not only in respect to safety of food of animal origin, but in general - requires a high degree of responsibility of people or organisations involved.

a)   Safety requires that the conditions of approval for a product remain valid as a result of the maintenance of product quality. This is the responsibility of the producer as well as it is his responsibility to supply all the data when submitting an application and to comply with the intentions of legislation in respect to the marketing of drugs.

b)   Safety requires that a product in fact is applied according to the conditions of use laid down on the approval. This is the responsibility of the veterinary profession and the farmer. Irresponsibility at that level has led to most of the so-called "drug scandals" in the past. To overcome this, cooperation between the two professions has to be improved. The use of more drugs does not necessarily bring about a better yield of production. The key to success rather seems to be the selective use of efficacious drugs based on a good diagnosis. However, it has to be recognized that the temptation for fraud will persist if no authorized drugs are available for important indications or if the conditions of use laid out are devoid of any practicability.

c)   Safety requires that a drug is rigorously checked on approval. The mechanisms to do so have been developed. To apply them adequately is the responsibility of the administrative body. Safety in respect to public health may be hampered not only by disparaging the one or the other potential risk, but also by the opposite attitude. As a result of an ultra or exaggerated sense of duty, unrealistic or unnecessary conditions of use may be laid down for a product which will again provoke fraud and hence the general risks associated with fraud.

d)   Safety requires the implementation of legislation and conditions of use. This is of utmost importance because it should not be forgotten that the consumer is generally not in a position to avoid residues if these are present in food. Implementation is primarily the responsibility of the executive organs. Yet it would be unfair to limit this aspect of safety to the executive. Particularly here we are confronted with a multi-factorial problem. By covering important other aspects like development of adequate analytical methods and of monitoring schedules, the other participants like the pharmaceutical industry or the veterinary profession are addressed here.

To conclude, I would like to state that a big step forward has been made in the past few years to guarantee the safety of food of animal origin in respect to the use of veterinary medicinal products. The present EC legislation seems to be a frame to adequately handle all pharmacologically active drugs intended for use in animals. Further diversifications would complicate the system and this should be avoided. An optimal safety will only be achieved if responsibility is an integral part of all actions taken. This also includes the part the consumer has to contribute. The system of feedback mechanisms will only work adequately if radicalism bare of any scientific basis is left out or - at least - remains a single event.

RESUME

LES OBJECTIFS DES LEGISLATIONS SUR LES PRODUITS VETERINAIRES

Bernd Hoffmann

Ambulatorische und Geburtshilfliche Veterinärklinik

Justus-Liebig-Universität-Giessen, 6300 Giessen

Le traitement, la prévention des maladies et l'amélioration des productions animales sont les principales indications d'utilisation des médicaments vétérinaires. Les producteurs (industrie pharmaceutique) et utilisateurs (fermiers) sont impliqués dans cette utilisation; les consommateurs la subissent du fait de la présence de résidus dans les denrées alimentaires d'origine animale. Comme dans tout système, des interactions vont se développer ou se sont développées. Livrés à eux-mêmes, les facteurs de contrôle qui sont évidents vont dominer, comme notamment l'efficacité ou le coût de l'utilisation.

Les éléments cachés a priori, tels que la sécurité des denrés d'origine animale sont moins évidents. C'est ici que la législation intervient pour garantir que les droits de chacun soient également respectés et pour qu'un équilibre s'installe. La législation européenne actuelle sur les médicaments vétérinaires remplit bien cet office. Un produit ne peut être approuvé que si les éléments de qualité, d'efficacité, de sécurité (tolérance) d'emploi pour l'animal et de sécurité pour la santé publique, ainsi que les conditions d'utilisation sont suffisamment documentés. Néanmoins, une telle législation est vouée à l'échec si toutes les personnes concernées ou intéressées ne se montrent pas suffisamment responsables. Ceci s'applique tant au fabricant et à l'utilisateur des produits vétérinaires, qu'aux organes exécutifs concernés et qu'aux consommateurs.

ZUSAMMENFASSUNG

VETERINÄR-ARZNEIMITTEL - ZIELE DER GESETZGEBUNG
Bernt Hoffmann,
Ambulatorische und Geburtshilfliche Veterinärklinik,
Justus-Liebig-Universität Gießen, 6300 Gießen.

Therapie, Krankheitsvorbeugung und verbesserte Produktion sind Hauptindikation für den Einsatz veterinärmedizinischer Produkte. Betroffen davon ist der Erzeuger (pharmazeutische Industrie) und der Gebraucher (Landwirt) und beeinflußt der Verbraucher infolge der Wahrscheinlichkeit von Rückständen in Nahrungsmitteln tierischen Ursprungs. Wie in jedem anderen System entwickelt sich getrennte Wege der Wechselwirkung und Feedback-Mechanismen, oder sie haben sich bereits entwickelt. Wenn man sie sich selbst überläßt, werden die unverkennbaren Kontrollfaktoren wie Wirksamkeit oder Kosten zu beherrschenden Faktoren.

Die "a priori" verborgenen Faktoren wie die Sicherheit der Nahrungsmittel tierischen Ursprungs wären weniger hervorstechend. In dieser Beziehung haben die Rechtsvorschriften zu gewährleisten, daß die legitimen Rechter aller Betroffenen gleich gut gewahrt werden, denn das System muß ausgeglichen sein. Die derzeitigen EG-Rechtsvorschriften in bezug auf veterinärmedizinische Produkte bieten einen angemessenen Rahmen. Ein Erzeugnis darf genehmigt werden wenn Qualität, Wirksamkeit, Sicherheit (Verträglichkeit) für das Tier und Sicherheit für die öffentliche Gesundheit, einschließlich der festgelegten Einsatzbedingungen im angemessenen Umfang nachgewiesen sind. Die Rechtsvorschriften in bezug auf Sicherheit werden jedoch nur dann Erfolg haben, wenn die betroffenen oder davon berührten Leute das erforderliche Verantwortungsbewußtsein haben. Das gilt sowohl für den Hersteller als auch für den Gebraucher veterinärmedizinischer Produkte, für die Exekutivorgane und den Verbraucher.

WHAT ARE THE OBJECTIVES OF THE LEGISLATION ON FEEDS AND FEED ADDITIVES

Dr. Gerhard Behm,

F.E.F.A.N.A.

Bonn 1

The legislation in the European Communities comprises a series of Directives which are either promulgated by the Council of Ministers of by the Commission. These Council or Commission Directives are not immediately applicable in the Member States, but must first be included in the national law on feeds and feed additives. The E.C. Directives thus only become effective after they have been incorporated in national legislation and promulgated at national level. The E.C. Directives actually represent guidelines for the national legislative authorities and not directly for the feed manufacturers in the various countries.

The Directives aim at the harmonization of legislation amongst Member States. This harmonization either goes into the details or is limited to general provisions. These provisions can be mandatory : "Member States shall..." or optional : "Member States can...". Various conditions determine this. The activity that is being regulated is particularly important as well as the size of the differences and interpretations of the existing legislations in the Member States.

Since the Directives are voted according to the principle of unanimity, the harmonization of the legislation can only be successful if it fits in with the interests of all Member States. It can thus take several years before a directive is ready for implementation.

There are presently five directives relating to the feed industry, namely :
- the directive on the circulation of straight feedingstuffs,
- the directive on the circulation of compounded feedingstuffs,
- the directive on dangerous substances,
- the directive on particular products for animal nutrition,
- the directive on feed additives.

I shall not talk about the first two of these directives, as they are not relevant in the context of this symposium. I shall briefly refer to the directive on dangerous substances and to that on products for animal nutrition, and then speak in more detail about the directive on feed additives.

The Directive on particular products for animal nutrition concerns those products that, directly or indirectly, represent sources of proteins, namely protein-products of micro-organic origin, non-protein nitrogen chains and amino-acids. The regulation of these particular feedingstuffs by a directive is justified by the new processes used to obtain these which require an approval which must be standardized in the E.C. Member States. The general concept of this directive is mostly to have a tool for approval and not to have a means to regulate the circulation of feedingstuffs, as is the case for the other directives. Regulatory approval should guarantee that the correct usage of the feedingstuffs in question does not represent any risk for either human or animal health, that the quality of the animal products is not adversely affected by this usage and that they have no adverse effect on the environment. This directive also aims at ensuring that the quality of food from animal origin is not negatively affected and also at protecting human and animal health.

This is also true for the directive which limits the levels of undesirable substances and products in feedingstuffs, which could in summary be referred to as dangerous substances. This directive concerns the undesired substances or products which are often contained in feedingstuffs and which could be detrimental to animal – or because of their presence in animal products, to human health.

Since these substances cannot be completely eliminated, a maximum limit had to be defined to prevent any undesirable adverse effect. Maximum levels of such substances in straight or compounded feedingstuffs have thus been fixed and regulations have been set up to deal with cases of non-respect of these approved limits. For a range of substances, such as heavy metals, nitrites, mycotoxins, prussic acid, plant impurities, different limits have been fixed depending on whether they are found in straight feedingstuffs or compounded feedingstuffs.

I shall now come to the main point of my exposé, namely the directive on additives in feedingstuffs for animal nutrition, to which we usually refer as the feed additive directive.

Shortly after the constitution of the Common Market in 1957, the Commission in Brussels started to compare the regulations on feedingstuffs which applied in the various Member States in an endeavour to standardize them. It was felt to be particularly urgent to formulate the approval and use conditions of additives in all Member States so that production conditions at least were the same in this sector and the free circulation of food from animal origin amongst Member States was possible without restriction.

When for example a product for poultry was used in France but was not approved in Italy because the level of residues was unacceptable in the latter country, the Italians had to close their border and stop imports of French poultry meat in Italy. This was obviously not the right way of running a community. This was recognized very early by the E.C. Commission who set up an "Expert Committee for Feed Additives" which was in charge of standardizing the conditions for approval and use of feed additives. This committee has been working on this programme since that time and will continue to do so.

This directive concerns all additive products which have a favourable effect on feedingstuffs themselves or on animal production. To be used for animal nutrition, these products should not be effective in controlling, treating or preventing disease. The regulation foresees that only the substances approved in this directive may be incorporated in feedingstuffs and their inclusion in feeds must correspond to the provision laid out in the directive. For these substances to be approved, it must be shown that they have a favourable effect on the quality of feedingstuffs and/or on animal products. They may not cause any health hazard for either human or animal health, nor may they be harmful to consumers of animal products. Furthermore, they may not be reserved for the treatment or prevention of disease, nor be limited to veterinary or medical therapeutic use.

At the beginning of its efforts towards the harmonization of legislations on feed additives, the Expert Committee first drew up an inventory of available products, classified per animal species and dosage levels, including the withdrawal times in the various countries. This inventory was the first step towards harmonization, so that nothing could be omitted and nobody was put at a disadvantage.

The second step was to identify those substances which were used in all countries and on which they could agree rapidly, i.e. substances without problems. The smallest common denominator was then identified to serve as a starting point for further steps. The substances identified in this manner were then classified into various categories such as antibiotics, growth promoters, antioxidants, flavouring agents, coccidiostats, emulsifiers, colouring agents, trace elements, vitamins, etc. and a list was established which is now Annex I of the directive. This Annex I lists all the additives which can be used in the same conditions throughout the Community.

For the sake of completeness it must also be said that this Directive has an Annex II. In this Annex are listed the additives which have been approved in a number of countries, but for which unanimous agreement has not been reached. However, the countries can approve these products for local usage if they wish and products from animals fed with feeds containing those products can circulate within the E.C. without restriction. Of course, only those products which have been tested and proved to be safe for human and animal health are included on these lists.

The directive on feed additives was issued in 1970, after several years of preparatory work, and was adopted in the national legislation of the six Member States within the next two years. It is thus Law since 1972 in all Member States, including now the countries that joined at a later stage.

The directive consists of three sections which must be regarded as a whole, since each individual section does not stand on its own. These are the text of the directive plus two annexes. The text prescribes the conditions whereby a product can be considered as a feed additive, what the dosage in replacement versus straight feedingstuffs should be, the label requirements, the conditions for approving a feed additive or removing it from the Annexes.

The conditions of admission of a feed additive to the Annexes are presented in the text of the directive, which states that a substance can only be listed in the annexes if it has a favourable effect on feedingstuffs or animal products, if the approved dosage has no harmful effect on human or animal health, if it brings no decrease of the quality of food from animal origin, if the substance can be analysed in the feed, if the approved dosage is neither therapeutic nor prophylactic, and finally if the substance is not reserved for human or veterinary therapy.

The above points are fundamental for the approval of a feed additive. These conditions assume that a number of studies have been conducted. Should any of them not be met, the manufacturer has no chance of ever receiving approval for his product. If the prescribed analytical methods do not work, or if the results vary from country to country, the substance will not be approved even though it may be the best.

A questionnaire exists which gives details of the studies to be conducted. This questionnaire is used in all Member States. It helps evaluate how carefully the work was done to produce a satisfactory substance and how everything possible was done to ensure the safety of animals and human beings. Although it would be interesting to know a bit more of the details of this questionnaire, I shall not read it to you now. Anyone who is interested in seeing it can obtain it.

The studies to be conducted for answering the questions in this document are defined in the "guidelines for the evaluation of additives for animal feedingstuffs". If the manufacturer does not follow these guidelines, the results are not authenticated and the studies must be done again.

When all the trials have been conducted and evaluated, all the information is presented in a thick dossier (several volumes), arranged in chapters and submitted to the authorities of the country where the manufacturer is located or represented. The dossier is examined by the national authorities. Scientists and laboratory experts are requested to check the validity of the trials and their results. If these are incomplete or incorrect the studies must be carried out again. If anything is missing, more information is requested.

After successful evaluation of the dossier by the national authorities, it is passed on to the E.C. Commission in Brussels with a request for inclusion in the annexes of the directive. At the same time, each Member State receives a copy of the dossier with translations of the summaries of every chapter.

The request is then evaluated by the Expert Committee for Feed Additives at the Commission. From our experience, it hardly ever occurs that a dossier is approved straight away by this body and additional questions or requests are often formulated. It frequently happens that the manufacturer must conduct further studies and produce additional data to satisfy these experts. When unanimity is reached in the Expert Committee (often after many years), the dossier is submitted to the Standing Committee with a recommendation. The Standing Committee decides on the inclusion in the annexes.

The majority of additives are also subject to scientific analyses by the Scientific Committee for Animal Nutrition which consists of scientists specialised in areas such as pharmacology and toxicology, microbiology, animal health, biology, nutrition, environment and which is at the disposal of the Commission. This Committee must usually also give positive advice before approval is granted for a substance.

One can surely and rightly say that no trial and approval system for additives has ever been safer or better than this one. Naturally, this creates difficulties for the companies who wish to obtain approval for a substance, but the principal objective of this legislation is to protect human and animal health. Health hazards must be eliminated as much as possible to maintain the confidence of the market partner, the consumer of food from animal origin.

Following positive votes from the concerned committees, the product is accepted in the directive, but no company identification is made. The product is referred to as additive X with its chemical name. This, in our opinion, is a serious handicap for the companies who develop those products. Anyone else can produce the substance from any source of material, without submitting a dossier, and capture his share of the market. He can thus produce at a lower cost than the company who had to carry out all the development work and conduct the studies required to obtain regulatory approval.

This is potentially dangerous and we keep drawing attention to this in Brussels and at the national authorities' levels. The product is of course evaluated in kidneys and liver and approved for its positive properties. However, for second generation products, or copied products, no one can be absolutely confident that they are identically safe or identically efficacious. It is possible - and it has been confirmed in trials - that fermentation products can behave differently on different carriers or in different quantities even if the ingredient produced is the same. Some products, which can legally be found on the market because they are approved under the terms of the directive, could present a risk. It could also be possible that a second generation product which is a good performance promoter does not have the expected effect because the label claims are not met. Fermentation or synthetic products with similar chemical structures or physical properties may have significantly variable biological efficacies, which could simply be different from what was expected or even be nil.

This thesis must be supported and FEFANA does support it. The interest of the research-based industry is of course also vital in this. The production of a satisfactory registration dossier when developing a new product costs approximately 30 million DM ($ 15 million). It is an important investment which no one would make for the benefit of somebody else. FEFANA is thus in favour of amending the directive in this direction, so that company-related or identified products are approved and not just substances. Other companies should of course have a chance to have access to the market with their products, but they should also prove their products are similar. We are not in favour of monopolies, but want to avoid the risk of seeing badly copied products appearing on the market and also feel that the research based industry must be encouraged to pursue its research and development efforts.

In summary, the objective of the E.C. legislation on feeds and feed additives is to guarantee the safety of these for both animal and human health. It can be achieved by the Directives which have been prepared in Brussels as long as the problems, to which I have referred earlier, are taken into account.

New definitions must be agreed on in the near future, so that only those additives which have actually been tested and which are proved to be safe can be marketed. This means that anyone who wishes to offer an additive on the market must also submit a dossier which includes experimental evidence. The serious industry can no longer accept that its efforts for quality and human safety be annihilated by the marketing of less well developed or less pure products.

We are looking forward to seeing an effort of the E.C. Commission in this direction and to a satisfactory outcome of it.

RESUME

LES OBJECTIFS DE LA LEGISLATION SUR
LES ALIMENTS ET ADDITIFS ALIMENTAIRES

Gerhard BEHM,
F.E.F.A.N.A.,
D-5300 Bonn

L'inclusion d'additifs dans l'alimentation animale a fait l'objet d'une harmonisation dans le cadre de la directive promulguée le 23 novembre 1970 par le Conseil des Ministres des Communautés Européennes. Les deux annexes de cette directive définissent en détail les additifs autorisés et leurs modalités d'emploi (dose, espèces animales, délais d'attente avant l'abattage). Cette directive est régulièrement mise à jour par la Commission avec la collaboration d'experts nationaux qui sont à sa disposition et du Comité Scientifique de l'Alimentation Animale. A l'heure actuelle 44 amendements ont ainsi été apportés à cette directive, ce qui témoigne d'une adaptation permanente à l'évolution dans le domaine des additifs alimentaires pour animaux. Ces amendements couvrent non seulement l'approbation de nouveaux produits, mais également la revision des modalités d'emploi de certains produits déjà repris aux annexes ou même l'exclusion de produits. Grâce à la flexibilité du système, l'utilisation des additifs correspond aux derniers développements des connaissances scientifiques, afin de limiter l'intérêt de la création de marchés noirs ou gris. C'est également vrai pour les médicaments vétérinaires du groupe des coccidiostats qui sont réglementés par cette directive sur les additifs alimentaires et sont donc bien contrôlés - une situation qui ne devrait pas changer.

Le texte de la Directive a fait l'objet de deux amendements et est encore en cours de révision. Cette dernière révision concerne principalement la réglementation de la distribution de ces produits, les questions de la déclaration et la préparation de monographies pour les additifs, qui les identifieraient par rapport à leur fabricant. L'objectif de ces mesures est de garantir la sécurité de ces produits et de maintenir, voire d'augmenter, les performances des animaux pour le bénéfice du consommateur.

ZUSAMMENFASSUNG

DIE ZIELE DER GESETZGEBUNG FÜR FUTTER- UND FUTTERZUSATZMITTEL

Gerhard Behm,

FEFANA, 5300 Bonn

Der Einsatz von Zusatzstoffen in der Tierernährung ist durch eine Richtlinie harmonisiert worden, die vom Ministerrat der EG am 23. November 1970 verabschiedet worden ist. Die beiden Anhänge der Richtlinie bestimmen im einzelnen sehr genau, welche Zusatzstoffe in welcher Dosierung bei welchen Tierarten eingesetzt werden dürfen und welche Absetzfristen vor der Schlachtung gegebenenfalls einzuhalten sind. Bei der EG-Kommission in Brüssel ist man mit Hilfe der dafür aus den Mitgliedsländern zur Verfügung stehenden Sachverständigen und mit Hilfe des Wissenschaftlichen Futtermittelausschußes ständig darum bemüht, die Anhänge der Richtlinie zu aktualisieren. Bis jetzt sind die Anhänge insgesamt 44 Mal geändert worden, was beweist, daß auf dem Gebiet der Zusatzstoffe eine zügige Entwicklung im Gange ist. Dabei werden neue Produkte zugelassen, die Anwendung der bereits in den Anhängen befindlichen Stoffe wird laufend überprüft und manche werden aus der Richtlinie auch wieder entfernt. Durch diese sehr flexible System entspricht die Verwendung von Zusatzstoffen immer dem neusten Stand der wissenschaftlichen Erkenntnisse, so daß von keiner Seite Interesse an der Entstehung eines grauen oder gar schwarzen Marktes bestehen kann. Das gilt auch für einige Arzneimittel aus der Gruppe der Kokzidiostatika, die über die Zusatzstoffrichtlinie geregelt und sehr gut kontrolliert eingesetzt werden, woran auch in Zukunft nichts geändert werden darf.

Der Textteil der Richtlinie ist bisher zweimal durch Änderungen verbessert worden und befindet sich zur Zeit in einer erneuten Überprüfung. Dabei geht es vorrangig um neue Verkehrsvorschriften für Zusatzstoffe, um Fragen der Deklaration und um die Erstellung von Monographien für Zusatzstoffe, damit die Identität der Produkte auf dem Markt überwacht und der Bezug zu den Herstellern geschaffen werden kann. Ziel all dieser Maßnahmen ist die Gewähr für Sicherheit beim Einsatz von Zusatzstoffen und die Erhaltung und Steigerung der Leistungsfähigkeit der Tiere zum Nutzen aller Verbraucher.

THE LIMITS OF LEGISLATION

D. Rees,

Lilly Research Centre, Windlesham, U.K.

There is a story that late in the day on September 22nd 1940 Winston Churchill, as Prime Minister of Great Britain at the time, received a telephone call from President Roosevelt to tell him the Americans had irrefutable evidence that the invasion of England would commence on September 23rd at 3 o'clock in the afternoon. The story goes that on the morning of September 23rd Churchill telephoned his Secretary of State for War, Anthony Eden, who was staying at his country home on the Kent coast at the time, and asked him to "walk down to the cliffs, to see if anyone was coming".

I invite you to join me in a walk 'to the cliffs' because there are a number of EEC issues approaching the shores of industry which command our attention.

The first issue that catches the eye is the Hormone situation. You will know that the Prohibition Directive of 1981 allowed for the review of five substances currently used for animal growth promotion in some of the Member States, and that three of these substances have since been cleared by a special Scientific Group. This same Scientific Group is continuing its appraisal of the remaining two substances on the basis of more recent data requested. In spite of this, there are continuing demands from Consumer Lobbyists that the use of hormones should be prohibited and, indeed, press reports indicate that the best proposal we can expect from the Commission regarding the two substances is that if they are not approved at a political level by a certain date, they WILL be banned for growth promotion use. If this report is correct then it opens up a new concept for the approval of substances and a very dangerous one. Until now it has been the long established policy of regulatory authorities to consider the authorisation of substances and products on the basis of scientific appraisal. It is on this basis that Industry researches its products and submits volumes of data to the regulatory authorities. If we are now to be faced with a new system which establishes authorisations on the basis of decisions made by politicians by an arbitrary date, then someone had better draw up new guidelines for Industry.

Now I know it can be argued that hormones are a special case and they <u>are</u> a political issue. But Industry has always held the view, and must continue to maintain the view, that the authorisation of their products must be based solely upon scientific appraisal, and that the political masters should arrive at their decisions on that basis. The Commission proposal - which has still to be published - has a long way to go yet, and there will be continuing pressures from the Consumerist Lobby for the prohibition of all hormones for growth promotion use. It is interesting to note that a few months ago Consumerists were stating there was no evidence to support the claim that hormones contribute at all to animal growth promotion. And yet, according to press reports, at a conference a few weeks ago they appeared to support a claim that the use of hormones accounts for 50 % of the Community meat stocks.. The whole demand for prohibition is based upon such shifting, shallow arguments. It is worth recalling at this stage that the hormone 'Scandal' of 1980 arose through the supply and use of DES in Member States where that substance was prohibited. In other words, it was supplied through a black market and use illegally which clearly indicates that prohibiting hormones will <u>not</u> stop their use.

I would agree that prohibition may sweep the whole matter under the carpet until another 'scandal' hits the public eye. But, as some enlightened members of Consumer Associations have recognised, it will do nothing to prevent a flourishing black market, from which misuse naturally follows and public safety is threatened. The grave danger is that if the lobbyists for prohibition are successful, a controllable situation will become uncontrollable and all future technical advances in this field will cease. And who knows ; when the Great Day comes that we have a sensible Agricultural Policy, those technical advances may well be needed.

There are two other issues on the horizon which are worth looking at: the proposal for a Medicated Feedingstuffs Directive and the possibility of a list of pharmacological molecules ; that is, a list of active substances for medicated premixes, which is referred to in the Veterinary Products Directive. As far as the Medicated Feedingstuffs Directive is concerned in my opinion it is questionable whether this is actually required at all, for there is no evidence that the intra-Community trade in Medicated Feedingstuffs is substantial enough _itself_ to require such a Directive. However, it may be that the feed industry feels the need to have Community standards laid down for the manufacture of these feeds. In this case medicated _premix_ manufacturers can hardly object as it is obviously in their own interests that their products should be used correctly in feedingstuffs. In addition, the proposal includes provisions for 'standard prescriptions' at both National and Community levels. My understanding is that 'standard prescriptions' (which I suggest would be better termed 'standard formulations') would enable feed compounders to maintain stocks of medicated feed which, of course, will permit economies of production - a distinct advantage for some Member States. So there appears to be two good reasons for having a directive. But the main argument against the proposal as it stands is that it provides for _veterinary prescription_ of _all_ medicated feeds. Whilst no one would suggest that all medicated feeds should be supplied free of veterinary control, it does not necessarily follow that all medicated feeds should require prescription. It is on this basis that the feed additive industry feels very strongly that the provision requiring veterinary prescription should be deleted from this proposal and the Member State authorities themselves should be allowed to decide which medicated feeds should be prescribed by the veterinarian.

As far as the list of pharmacological molecules is concerned, there are a number of objections to this concept. First of all, it is the premix preparation which is added to the feed - not simply the active ingredient. Therefore it is the premix - the preparation itself - which requires authorisation, and this is already covered by the veterinary products directive. Secondly, such a list implies a two-tier registration procedure - first the active substance, then the preparation. This would involve an _additional_ Community procedure which could only place further burdens on premix manufacturers for no good reasons, and considerably extend the time taken to make the product available to the market. Thirdly, there is a very real fear that the objective of having such a list is to severely restrict the number of substances used for medicating feedingstuffs.

This may seem strange at a time when the EEC Commission is spending millions of ECU's to encourage research and development in a number of fields - it certainly seems strange to me - but it is difficult to see any other purpose behind such a list. With that thought in mind I think it is worth reminding ourselves of Article 39 of the Treaty, dealing with the objectives of the Common Agricultural Policy. Under paragraph A, one of the objectives laid down is to "increase agricultural productivity by <u>promoting</u> technical progress". I repeat 'by <u>promoting</u> technical progress' - not by killing it off.

The fourth and last objection I will put to you today is that with the provisions for Community 'standard prescriptions' (or 'standard formulations') - as proposed in the Medicated Feedstuffs Directive - a list of molecules becomes unnecessary. This 'standard formulation' - that is ; the complete formulation for the medicated feed - will be controlled by the Standing Veterinary Committee and published in the Official Journal so that, in fact, a Community list will be established in time. Surely <u>this</u> list would make far more sense than a list of molecules.

I would now like to turn attention to another issue which we can view from our clifftop. One which has been 'at sea' for 6,5 years now. It is the proposal for a third amendment to the Directive concerning Additives in Feedingstuffs.

Now, many years ago - before it became as obvious as it is now - the Commission recognised a basic problem inherent in listing substances in the Annexes to the Directive, as opposed to preparations. This related to the problem of ensuring that substances which had lost patent protection and were being produced and marketed by 'secondary' manufacturers met the same standards of safety and efficiency as the substances approved. This is very much a question of public health, because no one knows, as there is no current system in most Member States to check, whether these secondary products meet the safety standards established. This means that the human food chain is exposed to substances which are of questionable standards.

The Commission therefore included a provision in the third amendment which provided for the listing of authorised producers as an Annex to the Directive. This meant that the primary producer, the innovator of the substance in most cases, who had submitted the original data, would be listed as an authorised producer and subsequent secondary producers would be listed as and when they showed that the substance they manufactured met the same standards as the substance listed. Now, one would have thought this proposal was a realistic, practical, even an obvious means of dealing with a real problem affecting public health. But no ! It appears that after the Lawyers moved in the whole concept became illegal under Community Law.

As a result new concepts are being considered. Apparently these include the preparation of a monograph - a 'profile' - for each listed substance which will have to be prepared by the primary producer and, presumably, approved by the Committee procedure established by the Commission. As yet it is not at all clear what is to be done with the monograph - nor who is to do, whatever is to be done with it ! What seems to be quite sure at present is that it is proposed to give priority to monographs for <u>new</u> substances, while existing substances would be dealt with over a 4 year period. Now, one realises that this matter is still under discussion. But it must be said that this is a disturbing feature because the problem of secondary substances does not arise from new substances, which are protected by patents, but from currently listed substances for which patent protection no longer exists.

It would therefore appear that serious consideration has been, and is being given, to a system which will impose additional burdens on primary producers without addressing the real problem at all. Now, I wish to make it clear that I personally believe there is no substitute for the original Commission proposal - a list of authorised producers. But if we are to be landed with a Monograph system, would it not be sensible to consider a system which has a direct bearing on the problem ?

The problem arises from the fact that when the patent expires for an approved, listed, substance, secondary manufacturers can enter the market uncontrolled. I therefore suggest we should forget about new substances entirely. The primary producers should be asked to produce monographs for those of their substances currently listed which have lost patent protection. I would guess there are between 15 to 20 such substances at present. These monographs should be dealt with urgently by the Commission. Following this, primary producers should be required to prepare monographs for any of the remaining substances listed for which the patent is due to expire within, say, 18 months. This would mean that once the current problems are dealt with, a continuing procedure would exist to deal with problems anticipated in the near future - not 5 to 15 years ahead.

Now I realise the Directive itself does not and cannot accommodate the subject of patent rights. And I make no pretence of having any idea how the wording of a Directive would need to be framed in order to cover the procedure I have outlined. All I do know is that such a procedure would concentrate attention where it is most needed, and save the Commission and its Committees a great deal unnecessary work. But of course producing Monographs is not enough. As long as a Community list of authorised producers is unacceptable to control secondary producers, it becomes essential that the Member State authorities institute some kind of procedure for the <u>registration</u> of secondary products. This is fundamental - otherwise how will the authorities know what secondary products are in the market place ? But the ultimate decision regarding the marketing authorisations for secondary products must rest with the Commission and its Standing Committee. After all, the feed additive directive is not merely concerned with intra-Community trade in feed additives ; it is primarily concerned with intra-Community trade in livestock and meatstuffs treated with those additives. Therefore, all Member States should be involved in the authorisation of secondary substances - just as they are with the authorisation of primary substances.

There are a number of issues contained in the proposed third amendment which are of interest to feed additive manufacturers. Amongst these are :
- the definition of an additive - which appears to include the preparation itself,
- the confidentiality of data,
- the distribution of substances,
- the labelling of additives.

I do not have time to discuss all of these, but there is one issue which is worthy of special attention. It is the proposal to lay down Minimum Conditions which must be fulfilled by manufacturers of additives, premixtures and Compound Feedingstuffs. I am sure that no one involved in the feed and feed additive industries established within the Community would object to a proposal to establish such conditions for manufacture ; indeed, in some Member States they already exist. But the question of interest to us all should be this :

> How will these conditions be applied to manufacturers in Third Countries ?

> How, for example, are the authorities to ensure the conditions are satisfied by a secondary manufacturer of a feed additive in Eastern Europe ?

Until that question is resolved it seems to me that any conditions imposed on manufacturers within the Community would lead to distortion of Trade. But even more important, if there is any logic to establishing manufacturing standards at all, at the end of the day it must surely relate to safety of the product. Are we then to say that an additive produced within the Community must be safe, but it doesn't matter if it is produced outside the Community ? It is an interesting thought and one which I must leave you with.

RESUME

LES LIMITES DES LEGISLATIONS

David E. Rees
Eli Lilly Research Centre,
Windlesham, U.K.

Les législations actuelles des Communautés Européennes sont brièvement passées en revue, afin d'attirer l'attention sur quelques problèmes qui peuvent entraîner une menace pour la sécurité des consommateurs de denrées d'origine animale.

Un des ces points est la prohibition des hormones comme promoteurs de croissance. Il apparaît quelques contradictions dans les opinions exprimées par les défenseurs des consommateurs, ce qui renforce l'impression que le problème est plus politique que scientifique. De plus, l'interdiction d'emploi des hormones pour la promotion de croissance dans la Communauté Européenne risque fort d'entraîner un développement du marché noir. Peut-on vraiment espérer empêcher cette évolution de la situation ? Comment pourra-t-on alors garantir la sécurité du consommateur ?

La proposition d'une Directive sur les Aliments Médicamenteux est un autre aspect des législations qui offre matière à la critique, de même que la liste des substances autorisées pour l'inclusion dans les aliments médicamenteux. En effet, une des premières critiques concerne la prescription vétérinaire obligatoire pour les aliments médicamenteux. S'il est bien certain que tous les aliments médicamenteux ne peuvent pas être fournis sans prescription vétérinaire, il n'est pas moins vrai par ailleurs que tous les aliments médicamenteux ne doivent pas nécessairement faire l'objet d'une telle prescription. La nécessité d'une prescription vétérinaire devrait être laissée à l'appréciation de chaque Etat Membre. En ce qui concerne la liste de substances, l'objection est fondée sur le fait qu'elles auront déjà fait l'objet d'une procédure d'approbation en vertu de la Directive sur les Médicaments Vétérinaires. Une telle liste imposerait à l'industrie une charge administrative supplémentaire. Finalement, cette proposition risque de limiter le nombre de substances autorisées. Ceci est en contradiction avec l'article 39 alinéa A du Traité de Rome qui reprend parmi les objectifs de la Politique Agricole Commune "l'accroissement de la productivité agricole par la promotion du progrès technique".

Enfin, les Annexes de la Directive sur les Additifs pour la Nutrition Animale ne reprennent que les listes de produits autorisés, classés sous forme de substances chimiques, sans relation avec le producteur initial. Une proposition avait été faite de reprendre dans les Annexes la liste des producteurs autorisés à fabriquer les substances en question et qui avaient soumis un dossier d'enregistrement. Les producteurs suivants figureraient sur cette liste après avoir soumis un dossier prouvant que les caractéristiques de leurs produits correspondent à celles des produits approuvés. Ceci aurait permis de s'assurer que les produits de la seconde génération offrent les mêmes garanties de sécurité et d'efficacité que les produits originaux dont le brevet aurait expiré. Toutefois, ce concept a été jugé illégal par les juristes du Marché Commun. D'autres propositions ont été soumises, comme par exemple les monographies (des profils) de chaque substance. Mais les modalités pour la soumission et l'approbation de ces monographies n'ont pas encore été proposées. De plus, il ne suffit pas d'établir des monographies pour de nouvelles substances; encore faut-il que les autorités puissent savoir que des produits de seconde génération sont sur le marché. Il ne s'agit pas uniquement ici de la protection des intérêts des producteurs originaux, mais aussi de la protection de la santé des consommateurs.

ZUSAMMENFASSUNG

DIE GRENZEN DER GESETZGEBUNG
David E. Rees,
Eli Lilly Research Centre,
Windlesham, G.B.

Vorgestellt wird eine kurze Übersicht derzeitiger Rechtsvorschriften in bezug auf Heilmittel und Wachstumsförderer für Tiere, wobei man auf eine Reihe von Fragen hinweist, die schlußendlich eine potentielle Bedrohung der Sicherheit des Verbrauchers von Nahrungsmitteln tierischen Ursprungs darstellen könnten.

Zu diesen Fragen gehört auch das Verbot von Hormonen als Wachstumsförderer. Bei den Meinungen, die von Verbraucher-Lobbyisten zum Ausdruck gebracht wurden, haben sich einige Widersprüche ergeben, die den Eindruck verstärken, hier handele es sich mehr um eine politische als um eine wissenschaftliche Frage. Außerdem dürfte das Verbot von Hormonen für die Wachstumsförderung in der EG wohl zu einem schwarzen Markt für Hormonen führen. Kann man dies bei realischer Betrachtungsweise verhindern ? Wie läßt sich die Sicherheit des Verbrauchers also gewährleisten ?

Vorschlag einer Richtlinie für arzneilich behandelte Futtermittel und Möglichkeit einer Liste aktiver Substanzen für arzneilich behandelte Vormischungen (pharmakologische Moleküle). Die Haupteinwände, die gegen den vorliegenden Vorschlag erhoben werden, betreffen in erster Linie die Bedingung eines tierärztlichen Rezeptes für alle arzneilich behandelten Futtermittel. Zwar empfiehlt niemand, alle arzneilich behandelten Futtermittel sollten außerhalb einer tierärztlichen Kontrolle geliefert werden, aber das bedeutet nicht unbedingt, daß jedes arzneilich behandelte Futtermittel rezeptpflichtig sein soll. Die Frage der Rezeptpflichtigkeit sollte der Entscheidung der einzelnen Mitgliedstaaten überlassen werden. Ein weiterer Einwand betrifft die Liste der Substanzen, die in arzneilich behandelten Vormischungen verwendet werden sollen. Diese Substanzen unterliegen jedoch bereits der Richtlinie über Veterinärerzeugnisse. Die vorgeschlagene Liste würde infolgedessen den Herstellern von Vormischungen eine zusätzliche verwaltungsmäßige Belastung aufbürden. Schließlich dürfte dieser Vorschlag die Zahl der Substanzen einschränken, die für arzneilich behandelte Futtermittel zur Verfügung stehen, was im Widerspruch zu den Bestimmungen des Art. 39, Abs, A des Vertrags zu stehen scheint, denn dieser Artikel befasst sich mit den Zielen der Gemeinsamen Agrarpolitik und eines dieser Ziele besteht darin, "die landwirtschaftliche Produktivität durch Förderung des technischen Fortschritts zu steigern".

Zum Schluß sei erwähnt, daß die Anlagen der Richtlinie Futterzusatzstoffe nur die Substanzen erwähnen, die als Zusatzstoffe für Futtermittel

genehmigt worden sind, ohne Bezugnahme auf den Hersteller. Es wurde vorge-
schlagen, man sollte die Primärerzeuger angeben, die die ursprünglichen Daten
eingereicht haben und spätere Sekundärerzeuger sollten dann in die Liste aufgenommen werden, wenn sie den Nachweis erbracht haben, daß die von ihnen her-
gestellte Substanz den gleichen Normen wie die in das Verzeichnis aufgenommene
Substanz entspricht. Damit hätte man gewährleisten können, daß Substanzen,
die ihren patentrechtlichen Schutz eingebüßt haben und von Sekundärerzeugern
hergestellt und vermarktet werden, den gleichen Normen in bezug auf Sicherheit
und Wirksamkeit wie das ursprüngliche Erzeugnis entsprechen. Von den Anwälten
der EG wurde dieses Konzept jedoch nach dem Gemeinschaftsrecht als rechts-
widrig erachtet. Es wurden weitere Konzepte in Betracht gezogen wie, bei-
spielsweise, Monographien - ein Profil - für jede Substanz. Hinsichtlich des
Einreichens oder Genehmigens solcher Monographien wurde jedoch noch kein Verfahren ausgearbeitet. Außerdem reicht es nicht, Monographien für neue oder
neuartige Substanzen zu erzeugen. Die Behörden sollten über ein Mittel ver-
fügen, durch das sie wissen, daß Sekundärerzeugnisse auf dem Markt sind, Es
geht nicht nur um den Schutz des Primärerzeuger, sondern auch um den Schutz
der Gesundheit des Verbrauchers.

CORRECT DRUG USAGE

William G. D. Montagu,
Crown Chemical Company Limited,
Lamberhurst, Kent, U.K.

This is a very large subject, or really two very large subjects. Firstly, what is a correct drug and secondly what do we mean by correct usage ? In the twenty minutes allotted to me, there will only be time to review the matters and to suggest policies that might serve the best interests of consumers and users.

So, let us start by attempting to define correct drugs and, therefore, the converse which one might call incorrect drugs.

Correct drugs could be considered to be those which are produced by manufacturers who have submitted satisfactory evidence as to the safety and efficacy of their product and its formulations to competent national or international authorities, and as a result have received authorisation for their sale and distribution through appropriate channels. Everything else must, by definition, be more or less incorrect which is where the problem starts.

Let us first consider the second generation drugs, sometimes called generics. These fall into two groups : those that have been subjected to the same searching investigations to establish safety and efficacy as the originals and those which have not; those which may be properly classified as "correct drugs" but those which have only been tested to establish conformity with some standard such as a Pharmacopoeia or a monograph may not be.

The active ingredient may have been produced or formulated in a different way which can affect the bioavailability, the carrier or base can affect excretion rates and residues. Oxytetracycline provides a good example.

We have heard from Dr. Marsboom how much effort and finance is needed to develop and test a new compound. Correct drugs are expensive. The second generation producer does not have to bear the development costs; but he should have to demonstrate that his product is in all ways equivalent, particularly bioequivalent, or if it is not that it is also safe and effective and I suggest that he should not be allowed to put it onto the market until he has done so.

Only in this way can the consumer and the user be safeguarded. Clear and enforced legislation is necessary.

Finally, there is a real danger : the illegal, unlicensed producer, who buys the cheapest raw materials he can find, relies upon someone else to have put the right label on the containers for his quality control and in some cases omits to include an active ingredient at all !

You may consider this to be an overstatement, so I will quote a case. In Great Britain, the Pharmaceutical Society is responsible for bringing prosecutions in such matters. Recently they investigated the illegal manufacture and distribution of an injectable stated to contain two antibiotics and a sulphonamide. An independent analytical laboratory was retained to provide evidence for a prosecution. Their first assays were unsatisfactory, so further investigations were made. In the end, they concluded that there was no active ingredient present at all, so the Pharmaceutical Society was unable to bring a prosecution for illegal sale of antibiotics. And, as an example of how the devil looks after his own, the time for bringing a prosecution under The Sale of Goods Act had run out !

The conclusion must be clear. In fact we can deal with this problem ourselves. Veterinarians, feed compounders and farmers should only use licensed products from reputable houses and then only obtain them from trustworthy distributors. There is always the need to beware of the counterfeiter, it is a great deal easier to copy a label and pack than its contents.

<u>Now let us turn to correct usage</u> : If you will accept my definition of a correct drug it follows that only correct drugs can be correctly used, because they are the only ones for which proper evidence of safety and efficacy is available. They are the only formulations where the instructions for use have been approved by independent and critical registration authorities.

Any manufacturer can submit data for his product and obtain registration and this is just what every reputable manufacturer does because he cannot affort mistakes.

Here are two examples. The authorities in one country were concerned about the amount of substandard and ineffective product being incorporated into poultry feed, so they told the only licensed producer of this important drug in that country to mend his ways.

The manufacturer in question came to the conclusion that, as his product only had 20 % of the market, the rest being illegally obtained, the easiest course was to stop production and leave the problem in the hands of the authorities. He withdrew his product.

Another concerned suspected resistance to oxytetracycline when incorporated into feed. It transpired that the compounder had changed his source of oxytetracycline to a cheaper producer who used a different manufacturing process. Chemical assays showed that the oxytetracycline was there but,

because of the method of manufacture, it only had 20 % of the bioavailability
of the licensed product. The buyer had taken the view that oxytetracycline
was a chemical and from his point of view the best was the cheapest. But, as
so often happens in this life, you get what you pay for.

Correct drug usage does not depend upon the distribution channel nor
upon who prescribes. It depends upon the user, be he a veterinary surgeon, a
feedstuff compounder or a farmer following manufacturer's instructions and
satisfying himself that the instructions have been based upon adequate evidence generated by competent people and subjected to critical examination.

If this rule was invariably followed, there would be no problem. So
why is there a problem and how big is it ?

To take the second point first, there are many shades of grey between
white and black. The size of the problem varies from country to country
depending upon the situation which arises as a result of legislation and
custom. Broadly speaking it seems to me that the more restrictive the legislation, the blacker the market, and in consequence the lower the standard of
the products used.

Sources vary. Amongst others there are :
1.   the veterinary surgeon who prescribes or supplies without preliminary
diagnosis, or who is tempted to economise by reducing his drug bill;
2.   the pharmacist who supplies his farmer friends with prescription
medicines;
3.   the manufacturer or wholesaler who supplies people he should not;
4.   the importer who, quite legally, imports chemicals and turns them into
pharmaceuticals, but lacks the facilities necessary to prove the safety and
efficacy of his product in order to provide clear instructions for use;
5.   and, finally, the smuggler who takes advantage of the fact that a product is permitted in one country but not in another. Growth promoters and
some antibiotics are good examples.

All this adds up to big business. It is not easy to get figures
because those who take part do not go in for sales audits or often such nice-
ties as V.A.T., T.V.A. or invoices, but we are talking about a trade worth
around $ 300 million a year with a gross profit of 60-70 %.

Where does it all go and what is it composed of ? There are not
reliable figures for reasons already mentioned. However, according to a mixture of opinions from many sources, the average relative value of the black −
or grey − market would be around 25 % and in some countries could be as high
as 70 %. Feed additives form a large part of illegal sales. In some

countries the figures quoted included drugs obtained by veterinary surgeons from unauthorized sources for fiscal reasons; the quality of products obtained in this way should not be suspected.

In France a large proportion of the feed additive market is supplied by large feedstuffs companies who make up their own premixes and who tend to obtain their raw materials from a variety of sources. There are spot markets in Hamburg and Amsterdam and a number of dealers.

The United Kingdom black market mainly consists of products which are restricted to veterinary surgeons by law such as antibiotics and trenbolone acetate. There is also a small but highly profitable trade in chloramphenicol.

Before we turn to the causes which have created this state of affairs and what can be done about it, we might be wise to ask ourselves the question : what is the object of the exercise ?

Agriculture is a big and important industry - the biggest single industry in the world, the most important industry in the world, the very source of life.

It is also a highly scientific industry and while it is pleasant for people of my age to look back and remember times when the hedgerows were full of flowers and the stubbles full of partridges, we have to remember that those same fields are now producing about twice the weight of grain. Pigs are ready for slaughter in not much over half the time, cattle can reach 450 kilos in 10 months. The list is endless.

Pharmaceuticals play a large part in this endeavour. In the E.C. hormonal and antibacterial growth promoters alone contribute a net of $ 1220 million in terms of meat at farm gate value. They also reduce imports of products like soya, saving some $ 150 million a year. These are not advantages to be case aside lightly.

I have not attempted to evaluate the part played by products used for therapeutic purpose but it must be enormous.

In 1969 Lord Swann's committee recommended that the use of antibacterials both antibiotics and sulphonamides, whether used therapeutically, as preventives or as growth promoters in farm animals or birds, should be restricted because it was believed at that time that the practice would lead to the development of resistant bacteria which could be transmitted to humans who consumed the meat, milk or eggs. Further research appears to have shown that this is not the case.

Is it now time to correct what may have been an overreaction to the Swann report and to classify these useful drugs under new headings and to divide them into different groups ?

1.   Products and formulations for therapeutic purposes where diagnosis is a necessary preliminary to treatment.

2.   Management tools which are used routinely by farmers and which do not require preliminary diagnosis before they are administered.

I do not propose to suggest lists, that is the registration authorities and the words of the United Kingdom Medicines Act could provide a guideline to identify "those products which with reasonable safety" may be sold directly to the user.

Many farmers are not unaware of the problem and its implications. In many cases they are well qualified men, often with degrees in agriculture. Nor are they irresponsible. They read journals, they go to conferences. They learn that there are developments and discoveries which will enable them to run their farms to greater advantage. They want to take advantage of these scientific advances.

If those management tools, coccidiostats, antibacterial growth promoters, hormonal implants, etc., are available to them in properly licensed and approved forms (with clear instructions for use) at reasonable prices, they would, I believe, prefer to use them and the problems will be reduced to more manageable proportions.

If not, we will be plagued with an ever increasing supply of unlicensed, uncontrolled and in many cases substandard products.

Who suffers as a result of all this activity ?   The consumer is disadvantaged for two reasons :

1.   The incorrect drugs tend to be older, simple molecules which possess side effects eliminated in more recently developed compounds.

2.   Incorrect drugs or correct drugs misused will not produce optimum results, so the final product costs more.

If the producer is deprived of the tools, information and advice which would enable him to improve efficiency, he suffers.

The citizens of the E.C. suffer because they have to pick up the bill for all this muddle and reduced productivity in the end.

In conclusion, the views expressed in this paper are my own. Figures given are my best estimates, so if what I have said has ruffled some feathers, do not take my words as the considered opinion of the D.S.A.

RESUME

L'UTILISATION CORRECTE DES MEDICAMENTS

William G. D. Montagu,

Crown Chemical Company Limited,

Lamberhurst, Kent, U.K.

Lorsque j'ai accepté, il y a six mois, de parler de ce sujet, et par extension de son alternative mieux connue sous le vocable de "marché noir", je ne réalisais pas à quoi je m'étais engagé. Le sujet est effectivement tellement vaste qu'il pourrait occuper un symposium à lui seul !

Il existe un vrai marché noir, où des produits efficaces, mais plus généralement des produits de second choix ou même totalement inefficaces, sont vendus à des clients crédules, vétérinaires, marchands ou fermiers, par des distributeurs ambulants qui réalisent des marges bénéficiaires de 60 à 70 %.

Il existe aussi plusieurs nuances de gris ...

Celles-ci couvrent plusieurs types de produits, notamment :

1.   les produits de firmes renommées, approuvés et fabriqués selon les normes nationales ou internationales, mais qui arrivent sur le marché par des canaux de distribution illicites;

2.   les produits approuvés et dont l'emploi est autorisé dans un pays, et non dans un pays voisin, et qui passent la frontière en contrebande;

3.   les produits de fabricants de moindre crédit, qui achètent leurs matières premières auprès de sources douteuses et omettent souvent d'effectuer les procédures de contrôle de qualité adéquates;

4.   les produits venant de laboratoires clandestins où aucun essai n'est jamais fait, aucun nom ne figure sur l'étiquette et les ingrédients ne correspondent pas aux indications. Il est possible d'obtenir du DES de cette manière;

5.   les produits de contrefaçon emballés de manière à ressembler aux médicaments authorisés.

De même, ces produits arrivent chez l'utilisateur final par divers canaux, comme par exemple :

1.   le vétérinaire qui prescrit sans diagnostic préalable;

2.   le pharmacien ou tout autre distributeur autorisé qui délivre le produit sans ordonnance;

3.   le fournisseur agricole qui parvient à obtenir des médicaments à ne délivrer que sur ordonnance par un canal éthique parallèle;

4.   l'importateur non autorisé qui fournit le marché noir local ou le

fermier directement;

5.   le fermier qui s'offre le voyage jusqu'à un pays plus libéral et ramène des stocks de produits pour lui et ses amis.

Le cumul de ces types de produits et canaux de distribution illicites représente un chiffre d'affaires de plus de 150 millions de dollars aux prix de 1982 et procure de gros bénéfices à des gens qui risquent peu de se faire prendre ou dont la peine, s'ils sont pris, sera dérisoire.

Finalement, que représente tout ceci dans le cadre de "Manger mieux et à sa faim", le titre de ce symposium ? Quels sont les dangers ? Qui risque quoi ? Que peut-on faire pour réduire - ou mieux éliminer - ce trafic qui n'existe que pour des motifs économiques ?

ZUSAMMENFASSUNG

KORREKTE ANWENDUNG VON ARZNEIMITTELN

William G. D. Montagu,
Crown Chemical Company Limited,
Lamberhurst, Kent, G.B.

Als ich mich vor sechs Monaten verpflichtete, über dieses Thema und über sein Gegenstück, den "Schwarzmarkt", zu sprechen, war mir noch nicht klar, was ich mir da eingebrockt hatte. Es erwies sich in der Tat als ein sehr weites Thema, dem man ohne weiters ein eigenes Symposium widmen könnte !

Es existiert tatsächlich ein Schwarzmarkt, der wirksame oder unter Standard wirksame, in vielen Fällen auch völlig wirkungslose Produkte an leichtgläubig Kunden vertreibt. Dazu zählen Tierärzte, Kaufleute und Landwirte, die direkt von den Wagen der Händler kaufen, wobei diese Profite von 60 bis 70 % erzielen.

Es gibt auch verschiedene Grauschattierungen ...

Zunächst gibt es mehrere Typen von Produkten, darunter :

1. die Produkte renommierter Firmen, lizenziert und hergestellt in Übereinstimmung mit den entsprechenden nationalen oder internationalen Standards; sie gelangten auf irgendeine Weise in nichtautorisierte Vertriebssysteme;

2. Produkte, die in einem Land lizenziert und gestattet, in einem anderen aber nicht zugelassen sind und über die Grenzen geschmuggelt werden;

3. die Produkte weniger angesehener Betriebe, die einfachere Wege gehen und ihre Rohstoffe von fragwürdigen Quellen beziehen und die oft keine richtigen Qualitätskontrollen durchführen;

4. die Produkte heimlicher Laboratorien, in denen keinerlei Tests durchgeführt werden; auf den Etiketts finden sich keine Namen, und oft stimmt der Inhalt nicht mit der Paketaufschrift überein. Es ist immer noch möglich, von solchen Quellen DES zu beziehen;

5. nachgemachte Produkte in einer Aufmachung, die der zugelassener Medikamente täuschend ähnlich sieht.

Es gibt viele Kanäle, auf denen diese Stoffe den Endverbraucher erreichen. Hier seien nur einige wenige genannt :

1. der Tierarzt, der ohne vorangegangene Diagnose ein Arzneimittel verschreibt;

2. der Apotheker oder ein anderer gesetzlicher Verteiler, der Arzneimittel ohne Verschreibung ausgibt;

3. der Fachhandel, dem es leicht fällt, Arzneimittel mit Einschränkungsbestimmungen zu erhalten;

4. ein nicht autorisierter Importeur, der den lokalen Schwarzmarkt und direkt die Landwirte beliefert;

5. der "do-it-yourself" Landwirt, der eine Reise in ein Land mit weniger

strengen Bestimmungen unternimmt und sich und seine Freunde versorgt.

Alles zusammengenommen gibt das ein Riesengeschäft in der Größen-
ordnung von 150 Mio Dollar, ausgedrückt in den Preisen von 1982. Die
Verteiler machen fette Profite, tragen ein nur sehr geringes Risiko und
müssen im Falle, daß sie entdeckt werden, nur lächerliche Strafen gewärtigen.

Schließlich und endlich, was hat dies für Auswirkungen für den Titel
und das Ziel dieses Symposiums "Sichere Lebensmittel heute und morgen" ?
Welche Gefahren kommen auf uns zu ? Wer trägt das Risiko ?

Was kann getan werden, daß dieser Nebenhandel, der nur aus wirtschaftlichen Gründen existiert, reduziert oder gar unterbunden wird ?

<u>DISCUSSION</u>

Question : To assist correct usage should the E.C. follow F.D.A. In setting up bioequivalency guidelines ? Should E.C. provide therein separate guidelines for veterinary medicines in therapy, coccidiostats for prophylaxis and antibacterials for performance promotion ?

B. Hoffmann : As I said during my presentation I am not at all in favour of setting up separate guidelines for veterinary products. I am personally quite unhappy that, for example, the hormones have been taken out of the general E.C. veterinary guidelines and the Scientific Commission of the E.C. which dealt with the hormone question stated that these guidelines were adequate to deal with the hormones specifically. So I would recommend that there is no further diversification but that the veterinary guidelines as they stand up today, as they are on the table and as they will be transformed into national laws should be enough to deal with all pharmacologically active products used in or on animals as veterinary drugs.

Question : Do you consider all contaminated food having a no toxic level as safe ? What supplementary conditions must be met before you consider a food as harmless to human health ?

B. Hoffmann : The question really relates to the residues in the foodstuff and indeed if the toxicological evaluation of the product leads to the establishment of a no effect - rather than no toxic - level in general, then we can come up with some further calculations to determine a level of residue concentrations in tissues which might be considered as safe. And, of course, this calculation includes well established additional safety factors as mentioned in the previous session. These safety factors are listed in a table and depend on the information available and on the type of drug. They range from 100 to 1000. In these conditions I think that we can define what is a harmless residue.

Question : When you speak about company related registration of additives, what do you mean ? Are you thinking of the many thousands of additives which are used now or of the small number that brings large profits to the pharmaceutical industry ?

G. Behm : I do not know what you mean with the thousands of additives, since we have only a limited number that is being used if I compare our Feed Additive Directive with the Feed Additive Compendium of the United States. Then I must say that we are particularly narrowhearted and have very few at our disposal. To answer the second part of your question, we envisage a company-related approval for additives only with respect to performance promoters and coccidiostats. These are products for which second or third manufacturers, who can obtain their raw materials from any mysterious source, can put a product on the market without any proof of quality. We would like to obtain that only proven additives for which companies assume the responsibility can be used and we know that this is also the wish of many national authorities and the E.C. Commission has shown understanding for our views. We can no longer assume responsibility for copy products which we cannot control.

Question : When after 10 years there is still no unanimous agreement in the E.C. about the products in Annex II, why is this Annex II not removed from the Directive ?

G. Behm : We think that Annex II fulfils an important function and precisely as introduction for new additives which should prove their value in practice before being approved in Annex I and authorised in all countries. We think that Annex II is also useful to remove a substance from the Directive. Annex I has no delay. Annex II has delays and a product is approved for a certain period of time. If this is not prolonged, the product leaves the Directive and is dead for further use. We thus need this Annex as an introductory stage or as exit for products. In addition there are situations when an additive is needed in several countries, but not in other states. This is also a reason for the need of Annex II, which gives the states the possibility to approve the products, but does not compel them to do so. For example : dimetridazole, ronidazole for pigs is necessary in pig farming in France and the U.K., but not in Germany.

Question : Coccidiostats do neither improve feed conversion nor weight gains. Why can coccidiostats not be registered as veterinary medicinals and be obtained without prescription ?

G. Behm : It is my personal opinion that coccidiostats belong to the feed additives directive and that we should not try to include them in the Medicated Feeds Directive, so as to include non-prescription products in this directive. It is written in this directive that any medicinal product for inclusion in the feeds must be prescribed by a veterinarian. We have two direc-

Dr. J. Boisseau, Directeur du LaboratoireNational des Médicaments Vétérinaires Fougères, France (discussion leader); Mr.David Miller, President of E.F.P.I.A. Working Party III (moderator).

Prof. René Truhaut, Laboratoire de Toxicologie, Faculté de Pharmacie, Paris.

tives : one for medicated feeds with veterinary prescription and one for additives without prescription. Thus if there are medicinal substances which can be distributed without prescription, they should go in the feed additives directive. This is why it exists. And if a substance must be distributed as a medicated feed, then please let us keep it in the Medicated Feeds Directive. Trying to include non prescription products in this directive would favour the creation of a new "grey" zone which we have tried to avoid.

Question : Who in your opinion will prepare and develop the particular monographs ?

D. Rees : I would imagine that the companies who submitted the dossiers by which the additive was listed in the annexes will be required to produce the monographs. As far as I know, anyway, discussions regarding the monograph concept are still continuing at Council experts' level and, therefore, no one has yet a clear picture as to how the system will operate. But I think certainly initially the company who submitted the dossier will have to prepare a monograph, a profile of that product for the authorities.

Question : You propose setting up a monograph when patents are finished. What would you do with products that are patented in one country and not in another and what do you do to all the feed grade products sold today ? Do you want a monograph for them too ?

D. Rees : I would like to try and deal with the second part of the question first. I assume that by feed grade products we are talking about the antioxidants and this type of additives. The proposed directive specifies monographs for antibiotics, coccidiostats and growth promoters only. If any of those products are not patented in all the countries, then I would have thought that the answer would be that in those countries that are patented would be the key for preparing the monographs. In other words if a product is patented say in Germany and France, but no other member state country, then within, I suggested an 18 months' period of the patent expiry, then monographs should be drawn up and submitted to the Commission for that product.

Intervention of Prof. René Truhaut.

In fact, while listening with much interest to the communications presented in this section, there is one particular point that retained my attention : the fact that from the time that a veterinary drug is marketed, based on the supply of adequate toxicological data, other companies which had not participated in the very important and very costly work required on the toxicological point of view, could benefit unduly from the investments made by the first companies. I shall take pesticides as an example, because in France we have been preoccupied by the same question, and as you may know we have a special commission to study the toxicity of antiparasitic and similar products used in agriculture. This commission must first give the "green light" before the homologation committee, which deals with the efficacy of the proposed agricultural products, can grant marketing approval. At the so-called Commission for the toxicology study, we have soon realised that the requirements we had, which become increasingly severe regarding the possibility of toxic effects for the consumers of treated crops because of the presence of residues, then for the health of those who manufacture or use those products, and lastly the absence of toxic effect of the compounds on the environment, i.e. on the aquatic, air or earth ecosystems. With the increase of requirements severity, the costs have also become heavier and you know how much such investments cost. So we thought that if we gave the green light, i.e. an approval for use, a favourable advice regarding the approval for use and if tomorrow or in the next few days, a company could copy the product, there was firstly a risk that the non respect of innovation would lead to a demotivation of the companies who had invested in research and saw others take all benefits from it, so there would be no interest in developing new molecules that could be useful to society, here to agriculture, and the protection of crops in general. Then there was also a more specifically toxicological preoccupation, namely that if a company could copy a product, there would no precisions regarding the composition of the product, we would not know what could vary, and consequently from the point of view of innocuousness disastrous consequences could be envisaged. As the Commission of Toxic Compounds is an Interministerial body which depends on the Ministry ofAgriculture and other Ministries, in particular the Ministry of Consumption, the toxicologists of this commission have something to do with consumer protection, which as I said is the final target of toxicologists who hope eventually to convince consumers that they are well protected. So what did we do ? We created an internal doctrine as follows : when we give an approval for an active substance which may be subject to a patent and which has not yet been used and which is presented in a suitable formulation for agriculture, we give approval and at the same time a 15 years' protection, i.e. any request for approval for a similar product by another company will have to be substantiated by a toxicological dossier as important as that asked from the first company. It is very logical on the toxicological point of view, but has provoked reactions at the level of the French Ministry of Economy and Finances. Inside that Ministry there is a Direction of Competition and I have been called on day by a triumvirat which told me that we were bringing a limiting factor to competition and in this manner we did not respect the interests of the users, because we gave monopoly to companies and did not permit to other companies to manufacture cheaper product. My answer was very short. I simply said : you do not know that plant health products, like many other products that may be a risk for public health, may not be treated from the point of view of competition like washing machines. If you want to limit the rule of 15 years that we have elaborated in full conscience of our responsibilities, I can tell you now that my colleagues and myself will react in such a way that the financial people will probably not be as well considered as they think they are. I do not know, Mr. President, if I have presented sufficiently well the policy applied in France, but which I think has been discussed at two FAO meetings last year and the year before, leading to recommendations in the same direction. I personally regret that up to now the Community has not dealt with these problems. It is disgraceful to permit that toxicological dossiers should be used - quite often badly - thus putting the population's health and the environment at risk, and should not correspond to the toxicological criteria which are legitimately required to approve the use of these products.

To be complete I shall add that when the 15 years are over, the second company that presents the product may use the toxicological data, but must supply identity and purity standards which ensure that the product is the same as that for which investigations were made.

CONCLUSIONS

1.   The programme of the symposium illustrates the strong wish of D.S.A. to have a constant interface with each of the sectors involved in the food chain. D.S.A. believes that this is the best way to achieve a good level of understanding and to develop systems which will satisfy everyone from the animal health industry to the consumer, including the veterinary profession and the farmer.

2.   The variety of interests represented in the attendance at the symposium confirms the need and desire for such interfaces between all the sectors involved in the food chain.

3.   It is recognized that the consumer organizations have an important role to play. Rather than basing studies and recommendations on emotional features however, it is suggested that they should adopt thorough scientific and technical approaches.

4.   D.S.A. believes that approval granted to market a product should be specific and not generic. D.S.A. is very concerned to see a legislative trend to protect generic products which in present conditions does not give any guarantee of safety to the users and the consumers.

5.   At present a list of substances is proposed in the Draft Directive on Medicated Feedingstuffs. D.S.A. together with other experts is opposed to this list which it believes is contrary to the Treaty of Rome and would represent a potential hazard to human health, because of lack of data on quality, safety and efficacy for substances on the list.

6.   A system of monographs is proposed in the Third Amendment of the Directive on Feed Additives. The system could be acceptable it if ensured that products were assessed according to these monographs before marketing. Unless there were detailed reviews of all new products coming to the market the monograph system could represent a danger to users and consumers.

7.   Due to continually increasing accuracy of analytical techniques, the attitude to the complete absence of detectable residues is no longer accept-

able. Expert opinion now considers that safe and tolerable residues provide acceptable standards for registration.

<u>RECOMMENDATIONS</u>

1.   Industry wishes to be consulted and advised at an early stage when legislation concerning its activities is in preparation. Regular communication would permit a full exchange of information and views with regulatory bodies.

2.   A satisfactory definition of the quality of food continues to be required. A more precise definition of quality can only be obtained as a result of a concentrated effort by all parties involved in the food chain. D.S.A. will make efforts to bring these parties together and agree parameters required.

3.   D.S.A. recommends that information on direction for use and warnings regarding the products reach every channel of distribution and product user. It is believed that this would help considerably to avoid improper use of products.

4.   D.S.A. believes that rigid and complex legislation on distribution channel is one important reason for product misuse and the development of a black market. It recommends that the responsibility to control the correct usage of animal health products should be left with individual countries according to a system which is sufficiently flexible and adapted to the nature of the market in each country.

5.   As the D.S.A. symposium has shown in the discussions on resistance issues, bacteriology is a science in constant evolution. It is recommended that a European information and investigation programme should be established to be shared between appropriate government departments and other experts, and the pharmaceutical industry. This would provide realistic data which would help in taking decisions based on facts and not on hypotheses.

6.   D.S.A. considers that productivity should not be confused with production. It believes that increased productivity through improved technology could solve the problems of under and overproduction. D.S.A. recommends that research and development into safe, advanced technology should be promoted for economical animal production.

7.   D.S.A. intends to act on the above recommendations and has established
a series of Working Parties to initiate the appropriate programmes. D.S.A.
invites all interested parties to present information to achieve these objectives and to assist in reaching the goals required.